STORM IN A TEACUP

STORM IN A TEACUP

The Physics of Everyday Life

HELEN CZERSKI

BANTAM PRESS

LONDON · NEW YORK · TORONTO · SYDNEY · AUCKLAND

TRANSWORLD PUBLISHERS
61–63 Uxbridge Road, London W5 5SA
www.penguin.co.uk

Transworld is part of the Penguin Random House group of companies
whose addresses can be found at global.penguinrandomhouse.com

First published in Great Britain in 2016 by Bantam Press
an imprint of Transworld Publishers

A CIP catalogue record for this book
is available from the British Library.

ISBNs 9780593075425 (cased)
9780593075432 (tpb)

Typeset in 11/14.5 pt Palatino by Jouve (UK), Milton Keynes
Printed and bound in Great Britain by Clays Ltd, Bungay, Suffolk

Penguin Random House is committed to a sustainable
future for our business, our readers and our planet. This book
is made from Forest Stewardship Council® certified paper.

1 3 5 7 9 10 8 6 4 2

To my parents,
Jan and Susan

While I was a university student, I spent a while doing physics revision at my Nana's house. Nana, a down-to-earth northerner, was very impressed when I told her that I was studying the structure of the atom.

'Ooh,' she said, 'and what can you do when you know that?'

It is a very good question.

Contents

Introduction

WE LIVE ON THE edge, perched on the boundary between planet Earth and the rest of the universe. On a clear night, anyone can admire the vast legions of bright stars, familiar and permanent, landmarks unique to our place in the cosmos. Every human civilization has seen the stars, but no one has touched them. Our home here on Earth is the opposite: messy, changeable, bursting with novelty and full of things that we touch and tweak every day. This is the place to look if you're interested in what makes the universe tick. The physical world is full of startling variety, caused by the same principles and the same atoms combining in different ways to produce a rich bounty of outcomes. But this diversity isn't random. Our world is full of patterns.

If you pour milk into your tea and give it a quick stir, you'll see a swirl, a spiral of two fluids circling each other while barely touching. In your teacup, the spiral lasts just a few seconds before the two liquids mix completely. But it was there for long enough to be seen, a brief reminder that liquids mix in beautiful swirling patterns and not by merging instantaneously. The same pattern can be seen in other places too, for the same reason. If you look down on the Earth from space, you will often see very similar swirls in the clouds, made where warm air and cold air waltz around each other instead of mixing directly. In Britain, these swirls come rolling across the Atlantic from the west on a regular basis, causing our notoriously changeable weather. They form at the boundary between cold polar air to the north and warm tropical air to the south. The cool and warm air chase each other around in circles, and you can see the pattern clearly on satellite images. We know these swirls as depressions or cyclones, and we experience rapid changes between wind, rain and sunshine as the arms of the spiral spin past.

A rotating storm might seem to have very little in common with a stirred mug of tea, but the similarity in the patterns is more than coincidence. It's a clue that hints at something more fundamental. Hidden beneath both is a systematic basis for all such formations, one discovered and explored and tested by rigorous experiments carried out by generations of humans. This process of discovery is science: the continual refinement and testing of our understanding, alongside the digging that reveals even more to be understood.

Sometimes a pattern is easy to spot in new places. But sometimes the connection goes a little bit deeper and so it's all the more satisfying when it finally emerges. For example, you might not think that scorpions and cyclists have much in common. But they both use the same scientific trick to survive, although in opposite ways.

A moonless night in the North American desert is cold and quiet. Finding anything out here seems close to impossible, since the ground is lit only by dim starlight. But to find one particular treasure, you equip yourself with a special torch and set out into the darkness. The torch needs to be one that produces light that is invisible to our species: ultraviolet light, or 'black light'. As the beam roams across the ground, it's impossible to tell exactly where it's pointing because it's invisible. Then there's a flash, and the darkness of the desert is punctured by a surprised scuttling patch of eerie bright blue-green. It's a scorpion.

This is how enthusiasts find scorpions. These black arachnids have pigments in their exoskeleton that take in ultraviolet light that we can't see and give back visible light that we can see. It's a

2

really clever technique, although if you're scared of scorpions to start with, your appreciation might be a little muted. The name for this trick of the light is fluorescence. The blue-green scorpion glow is thought to be an adaptation to help the scorpions find the best hiding places at dusk. Ultraviolet light is around all the time, but at dusk, when the sun has just slipped below the horizon, most of the visible light has gone and only the ultraviolet is left. So if the scorpion is out in the open, it will glow and be easy to spot because there isn't much other blue or green light around. If the scorpion is even slightly exposed, it can detect its own glow and so it knows it needs to do a better job of hiding. It's an elegant and effective signalling system – or was until the humans bearing ultraviolet torches turned up.

Fortunately for the arachnophobes, you don't need to be in a scorpion-populated desert at night to see fluorescence – it's pretty common on a dull morning in the city as well. Look again at those safety-conscious cyclists: their high-visibility jackets seem oddly bright compared with the surroundings. It looks as though they're glowing, and that's because they are. On cloudy days, the clouds block the visible light, but lots of ultraviolet still gets through. The pigments in the high-visibility jackets are taking in the ultraviolet and giving back visible light. It's exactly the same trick the scorpions are playing, but for the opposite reason. The cyclists *want* to glow; if they're emitting that extra light, they're easier to see and so safer. This sort of fluorescence is pretty much a free lunch for humans; we're not aware of the ultraviolet light in the first place, so we don't lose anything when it gets turned into something we can use.

It's fascinating that it happens at all, but the real joy for me is that a nugget of physics like that isn't just an interesting fact: it's a tool that you carry with you. It can be useful anywhere. In this case, the same bit of physics helps both scorpions and cyclists survive. It also makes tonic water glow under ultraviolet light, because the quinine in it is fluorescent. And it's how laundry brighteners and highlighter pens work their magic. Next time you look at a high-lighted paragraph, bear in mind that the highlighter ink is also acting as an ultraviolet detector; even though you can't see the ultraviolet directly, you know it's there because of this glow.

I studied physics because it explained things that I was inter-ested in. It allowed me to look around and see the mechanisms making our everyday world tick. Best of all, it let me work some of them out for myself. Even though I'm a professional physicist now, lots of the things I've worked out for myself haven't involved laboratories or complicated computer software or expensive experiments. The most satisfying discoveries have come from random things I was playing with when I wasn't meant to be doing science at all. Knowing about some basic bits of physics turns the world into a toybox.

There is sometimes a bit of snobbery about the science found in kitchens and gardens and city streets. It's seen as something to occupy children with, a trivial distraction which is important for the young, but of no real use to adults. An adult might buy a book about how the universe works, and that's seen as being a proper adult topic. But that attitude misses something very important: the same physics applies everywhere. A toaster can teach you about some of the most fundamental laws of physics, and the benefit of a toaster is that you've probably got one, and you can see it working for yourself. Physics is awesome precisely because the same patterns are universal: they exist both in the kitchen and in the furthest reaches of the universe. The advan-tage of looking at the toaster first is that even if you never get to worry about the temperature of the universe, you still know why your toast is hot. But once you're familiar with the pattern, you will recognize it in many other places, and some of those other places will be the most impressive achievements of human soci-ety. Learning the science of the everyday is a direct route to the background knowledge about the world that every citizen needs in order to participate fully in society.

Have you ever had to tell apart a raw egg and a boiled egg, without taking their shells off? There's an easy way to do it. Put the egg down on a smooth, hard surface and set it spinning. After a few seconds, briefly touch the outside of the shell with one finger, just enough to stop the egg's rotation. The egg might just sit there, stationary. But after a second or two, it might slowly start to spin again. Raw and boiled eggs look the same on the outside, but their insides are different and that gives the secret away. When you touched the cooked egg, you stopped a whole solid object. But when you stopped the raw egg, you only stopped the shell. The liquid inside never stopped swirling around, and so after a second or so, the shell started to rotate again, because it was being dragged around by its insides. If you don't believe me, go and find an egg and try it. It is a principle of physics that objects tend to continue the same sort of movement unless you push or pull on them. In this case, the total amount of spin of the egg white stays the same because it had no reason to change. This is known as conservation of angular momentum. And it doesn't just work in eggs.

The Hubble Space Telescope, an orbiting eye that has been whooshing round our planet since 1990, has produced many thousands of spectacular images of the cosmos. It has sent back pictures of Mars, the rings of Uranus, the oldest stars in the Milky Way, the wonderfully named Sombrero Galaxy and the giant Crab Nebula. But when you're floating freely in space, how do you hold your position as you gaze on such tiny pinpoints of light? How do you know precisely which way you're facing? Hubble has six gyroscopes, each of which is a wheel spinning at 19,200 revolutions per second. Conservation of angular momentum means that those wheels will keep spinning at that rate because there is nothing to slow them down. And the spin axis will stay pointed in precisely the same direction, because it has no reason to move. The gyroscopes give Hubble a reference direction, so that its optics can stay locked on a distant object for as long as necessary. The physical principle used to orient one of the most advanced technologies our civilization can produce can be demonstrated with an egg in your kitchen.

This is why I love physics. Everything you learn will come in useful somewhere else, and it's all one big adventure because you don't know where it will take you next. As far as we know, the physical laws we observe here on Earth apply everywhere in the universe. Many of the nuts and bolts of our universe are accessible to everyone. You can test them for yourself. What you can learn with an egg hatches into a principle that applies everywhere. You step outside armed with your hatchling, and the world looks different.

In the past, information was treasured more than it is now. Each nugget was hard-earned and valuable. These days, we live on the shore of an ocean of knowledge, one with regular tsunamis that threaten our sanity. If you can manage your life as you are, why seek more knowledge and therefore more complications? The Hubble Space Telescope is all very nice, but unless it's also going to look downwards once in a while to find your keys when you're late for a meeting, does it make any difference?

Humans are curious about the world, and we get a lot of joy from satisfying our curiosity. The process is even more rewarding if you work things out for yourself, or if you share the journey of discovery with others. And the physical principles you learn from playing also apply to new medical technologies, the weather, mobile phones, self-cleaning clothes and fusion reactors. Modern life is full of complex decisions: is it worth paying more for a compact fluorescent light bulb? Is it safe to sleep with my phone next to my bed? Should I trust the weather forecast? What difference does it make if my sunglasses have polarizing lenses? The basic principles alone often won't provide specific answers, but they'll provide the context needed to ask the right questions. And if we're used to working things out for ourselves, we won't feel helpless when the answer isn't obvious on the first try. We'll know that with a bit of extra thinking, we can clarify things. Critical thinking is essential to make sense of our world, especially with advertisers and politicians all telling us loudly that they know best. We need to be able to look at the evidence and work out whether we agree with them. And there's more than our own daily lives at stake. We are responsible for our civilization. We vote, we choose what to buy and how to live, and we are collectively part of the human journey. No one can understand every single detail of our complex world, but the basic principles are fantastically valuable tools to take with you on the way.

Because of all this, I think that playing with the physical toys in the world around us is more than 'just fun', even though I'm a huge fan of fun for its own sake. Science isn't just about collecting facts; it's a logical process for working things out. The point of science is that everyone can look at the data and come to a reasoned conclusion. At first, those conclusions may differ, but then you go and collect more data that helps you decide between one description of the world and another, and eventually the conclusions converge. This is what separates science from other disciplines – a scientific hypothesis must make specific testable predictions. That means that if you have an idea about how you think something works, the next thing to do is to work out what the consequences of your idea would be. In particular, you have to look hard for consequences that you can check for, and especially for consequences that you can prove wrong. If your hypothesis passes every test we can think of, we cautiously agree that this is probably a good model for the way the world works. Science is always trying to prove itself wrong, because that's the quickest route to finding out what's actually going on.

You don't have to be a qualified scientist to experiment with the world. Knowing some basic physical principles will set you on the right track to work a lot of things out for yourself. Sometimes, it doesn't even have to be an organized process – the jigsaw pieces almost slot themselves into place.

One of my favourite voyages of discovery started with disappointment: I made blueberry jam and it turned out pink. Bright fuchsia pink. It happened a few years ago, when I was living in Rhode Island, sorting out the last bits and pieces before moving back to the UK. Most things were done, but there was one last project that I was adamant about fitting in before I left. I had always loved blueberries – they were slightly exotic, delicious, and beautifully and bizarrely blue. In most places I've lived they come in frustratingly small quantities, but in Rhode Island they grow in abundance. I wanted to convert some of the summer blueberry bounty into blue jam to take back to the UK. So I spent one of my last mornings there picking and sorting blueberries.

The most important and exciting thing about blueberry jam is surely that it is blue. I thought so, anyway. But nature had other ideas. The pan of bubbling jam was many things, but blue was not one of them. I filled the jam jars, and the jam really did taste lovely. But the lingering disappointment and confusion followed me and my pink jam back to the UK.

Six months later, I was asked by a friend to help with a historical conundrum. He was making a TV programme about witches, and he said that there were records of 'wise women' boiling verbena petals in water and putting the resulting liquid on people's skin as a way of telling whether they were bewitched. He wondered whether they were measuring something systematically, even if it wasn't what they intended. I did a bit of research and found that maybe they were.

Purple verbena flowers, along with red cabbage, blood oranges and lots of other red and purple plants, contain chemical compounds called anthocyanins. These anthocyanins are pigments, and they give the plants their bright colours. There are a few different versions, so the colour varies a bit, but they all have a similar molecular structure. That's not all, though. The colour also depends on the acidity of the liquid that the molecule is in – what's called its 'pH value'. If you make that environment a little more acidic or a little more alkaline, the molecules change shape slightly and so their colour changes. They are indicators, nature's version of litmus paper.

You can have lots of fun in the kitchen with this. You need to boil the plant to get the pigment out, so boil a bit of red cabbage in water, and then save the water (which is now purple). Mix some with vinegar, and it goes red. A solution of laundry powder (a strong alkali) makes it go yellow or green. You can generate a whole rainbow of outcomes just from what's in your kitchen. I know: I did it. I love this discovery because these anthocyanins are everywhere, and accessible to anyone. No chemistry set required!

So maybe these wise women were using the verbena flowers to test for pH, not bewitchment. Your skin pH can vary naturally, and putting the verbena concoction on skin could produce different colours for different people. I could make cabbage water go from purple to blue when I was nice and sweaty after a long run, but it didn't change colour when I hadn't been exercising. The wise women may have noticed that different people made the verbena pigments change in different ways, and put their own interpretation on it. We'll never know for sure, but it seems to me to be a reasonable hypothesis.

So much for history. And then I remembered the blueberries and the jam. Blueberries are blue because they contain anthocyanins. Jam has only four ingredients: fruit, sugar, water and lemon juice. The lemon juice helps the natural pectin from the fruit do its job of making the jam set. It does that because . . . it's acid. My blueberry jam was pink because the boiled blueberries were acting as a saucepan-sized litmus test. It had to be pink for the jam to set properly. The excitement of working that out almost made up for the disappointment of never having made blue jam. Almost. But the discovery that there's a whole rainbow of colour to be had from just one fruit is the sort of treasure that's worth the sacrifice.

This book is about linking the little things we see every day with the big world we live in. It's a romp through the physical world, showing how playing with things like popcorn, coffee stains and refrigerator magnets can shed light on Scott's expeditions, medical tests and solving our future energy needs. Science is not about 'them', it's about 'us', and we can all go on this adventure in our own way. Each chapter begins with something small in the everyday world, something that we will have seen many times but may never have thought about. By the end of each chapter, we'll see the same patterns explaining some of the most important science and technology of our time. Each mini-quest is rewarding in itself, but the real payoff comes when the pieces are put together.

There's another benefit to knowing about how the world works, and it's one that scientists don't talk about often enough. Seeing what makes the world tick changes your perspective. The world is a mosaic of physical patterns, and once you're familiar with the basics, you start to see how those patterns fit together. I hope that as you read this book, the scientific hatchlings from the chapters along the way will grow into a different way of seeing the world. The final chapter of this book is an exploration of how the patterns interlock to form our three life-support systems – the human body, our planet and our civilization. But you don't have to agree with my perspective. The essence of science is experimenting with the principles for yourself, considering all the evidence available and then reaching your own conclusions.

The teacup is only the start.

I

Popcorn and rockets

The gas laws

EXPLOSIONS IN THE KITCHEN are generally considered a bad idea. But just occasionally, a small one can produce something delicious. A dried corn kernel contains lots of nice food-like components – carbohydrates, proteins, iron and potassium – but they're very densely packed and there's a tough armoured shell in the way. The potential is tantalizing, but to make it edible you need some extreme reorganization. An explosion is just the ticket, and very conveniently, this seed carries the seeds of its own destruction within it. Last night, I did a bit of ballistic cooking and made popcorn. It's always a relief to discover that a tough, unwelcoming exterior can conceal a softer inside – but why does this one make fluff instead of blowing itself to bits?

Once the oil in the pan was hot, I added a spoonful of kernels, put the lid on, and left it while I put the kettle on to make tea. Outside, a huge storm was raging, and chunky raindrops were hammering against the window. The corn sat in the oil and hissed gently. It looked to me as though nothing was happening, but inside the pan, the show had already started. Each corn kernel contains a germ, which is the start of a new plant, and the endosperm, which is there as food for the new plant. The endosperm is made up of starch packaged into granules, and it contains about 14 per cent water. As the kernels sat in the hot oil, that water was starting to evaporate, turning into steam. Hotter molecules move faster, so that as each kernel heated up, there were more and more water molecules whooshing around inside it as steam. The evolutionary purpose of a corn kernel's shell is to withstand assault from outside, but it now had to contain an internal rebellion – and it was acting like a mini pressure cooker. The water molecules that had turned to steam were trapped with nowhere to go, so the pressure inside was building up. Molecules of gas were continually bumping into each other and into the walls of the container, and as the number of gas molecules increased and they moved faster, they were hammering harder and harder on the inside of the shell.

Pressure cookers work because hot steam cooks things very effectively, and it's no different inside popcorn. As I searched for teabags, the starch granules were being cooked into a pressurized gelatinous goo, and the pressure kept going up. The outer shell of a popcorn kernel can withstand this stress, but only up to a point. When the temperature inside approaches 180°C and the pressure gets up to nearly ten times the normal pressure of the air around us, the goo is on the edge of victory.

I gave the pan a little shake and heard the first dull pop echoing round the inside. After a couple of seconds, it sounded as though a mini machine gun was being fired in there, and I could see the lid lifting as it got hit from underneath. Each individual pop also came with a fairly impressive puff of steam from the edge of the pan lid. I left it for a moment to pour a cup of tea, and in those few seconds, the barrage from underneath shifted the lid and fluff started taking flight.

At the moment of catastrophe, the rules change. Until that point, a fixed amount of water vapour is confined, and the pressure it exerts on the inside of the shell increases as the temperature increases. But when the hard shell finally succumbs, the insides are exposed to the atmospheric pressure in the rest of the pan and there is no volume limit any more. The starchy goo is still full of hot hammering molecules but nothing is pushing back from the other side. So it expands explosively, until the pressure inside matches the pressure outside. Compact white goo becomes expansive white fluffy foam, turning the entire kernel inside-out; and as it cools, it solidifies. The transformation is complete.

Tipping the popped corn out revealed a few casualties left behind. Dark burnt unpopped corn rattled sadly round the bottom of the pan. If the outer shell is damaged, water vapour escapes as it is heated, and the pressure never builds up. The reason that popcorn pops and other grains don't is that all the others have porous shells. If a kernel is too dry, perhaps because it was harvested at the wrong time, there isn't enough water inside it to build up the pressure needed to burst the shell. Without the violence of an explosion, inedible corn remains inedible.

I took the bowl of perfectly cooked fluff and the tea over to the window and stood watching the storm. Destruction doesn't always have to be a bad thing.

*

There is beauty in simplicity. And it's even more satisfying when that beauty condenses out of complexity. For me, the laws that tell us how gases behave are like one of those optical illusions where you think you're seeing one thing, and then you blink and look again and see something completely different.

We live in a world made of atoms. Each of these tiny specks of matter is coated with a distinctive pattern of negatively charged electrons, chaperones to the heavy and positively charged nucleus within. Chemistry is the story of those chaperones sharing duties between multiple atoms, shifting formation while always obeying the strict rules of the quantum world, and holding the captive nuclei in larger patterns called molecules. In the air I'm breathing as I type this, there are pairs of oxygen atoms (each pair is one oxygen molecule) moving at 900 mph bumping into pairs of nitrogen atoms going at 200 mph, and then maybe bouncing off a water molecule going at over 1,000 mph. It's horrifically messy and complicated – different atoms, different molecules, different speeds – and in each cubic centimetre of air there are about 30,000,000,000,000,000,000 (3×10^{19}) individual molecules, each colliding about a billion times a second. You might think that the sensible approach to all that is to quit while you're ahead and take up brain surgery or economic theory or hacking supercomputers instead. Something simpler, anyway. So it's probably just as well that the pioneers who discovered how gases behave had no idea about any of it. Ignorance has its uses. The idea of atoms wasn't really a part of science until the early 1800s and absolute proof of their existence didn't turn up until around 1905. Back in 1662, all that Robert Boyle and his assistant Robert Hooke had was glassware, mercury, some trapped air and just the right amount of ignorance. They found that as the pressure on a pocket of air increased, its volume decreased. This is Boyle's Law, and it says that gas pressure is inversely proportional to volume. A century later, Jacques Charles found that the volume of a gas is directly proportional to its temperature. If you double the temperature, you double the volume. It's almost unbelievable. How can so much atomic complication lead to something so simple and so consistent?

*

One last intake of air, one calm flick of its fleshy tail, and the giant leaves the atmosphere behind. Everything this sperm whale needs to live for the next forty-five minutes is stored in its body, and the hunt begins. The prize is a giant squid, a rubbery monster armed with tentacles, vicious suckers and a fearsome beak. To find its prey, the whale must venture deep into the real darkness of the ocean, to the places never touched by sunlight. Routine dives will reach 500–1,000 metres, and the measured record is around 2 km. The whale probes the blackness with highly directional sonar, waiting for the faint echo that suggests dinner might be close. And the giant squid floats unaware and unsuspecting, because it is deaf.

The most precious treasure the whale carries down into the gloom is oxygen, needed to sustain the chemical reactions that power the swimming muscles, and the whale's very life. But the gaseous oxygen supplied by the atmosphere becomes a liability in the deep – in fact, as soon as the whale leaves the surface, the air in its lungs becomes a problem. For every additional metre it swims downwards, the weight of one extra metre of water presses inwards. Nitrogen and oxygen molecules are bouncing off each other and the lung walls, and each collision provides a minuscule push. At the surface, the inward and outward pushes on the whale balance. But as the giant sinks, it is squashed by the additional weight of the water above it, and the push of the outside overwhelms the push from the inside. So the walls of the lungs move inwards until equilibrium, the point where the pushes are balanced once again. A balance is reached because as the whale's lung compresses, each of the molecules has less space and collisions between them become more common. That means that there are more molecules hammering outwards on each bit of the lungs, so the pressure inside increases until the hammering molecules can compete equally with those outside. Ten metres of water depth is enough to exert additional pressure equivalent to a whole extra atmosphere. So even at that depth, while it could still easily see the surface (if it were looking), the whale's lungs reduce to half the volume that they were. That means there are twice as many molecular collisions on the walls, matching the doubled pressure from outside. But the squid might be 1 km below the surface, and at that depth the vast pressure of the water means that the lungs should collapse to a mere 1 per cent of the volume they have at the surface.

Eventually, the whale hears the reflection of one of its loud clicks. With shrunken lungs, and only sonar to guide it, it must now prepare for battle in the vast darkness. The giant squid is armed, and even if it eventually succumbs, the whale may well swim away with horrific scars. Without oxygen from its lungs, how does it even have the energy to fight?

The problem of the shrunken lungs is that if their volume is only one-hundredth of what it was at the surface, the pressure of the gas in there will be one hundred times greater than atmospheric pressure. At the alveoli, the delicate part of the lungs where oxygen and carbon dioxide are exchanged into and out of the blood, this pressure would push both extra nitrogen and extra oxygen to dissolve in the whale's bloodstream. The result would be an extreme case of what divers call 'the bends', and as the whale returned to the surface the extra nitrogen would bubble up in its blood, doing all sorts of damage. The evolutionary solution is to shut off the alveoli completely, from the moment the whale leaves the surface. There is no alternative. But the whale can access its energy reserves because its blood and muscles can store an extraordinary amount of oxygen. A sperm whale has twice as much haemoglobin as a human, and about ten times as much myoglobin (the protein used to store energy in the muscles). While it was at the surface, the whale was recharging these vast reservoirs. Sperm whales are never breathing from their lungs when they make these deep dives. It's far too dangerous. And they're not just using their one last breath while they're underwater. They're living – and fighting – on the surplus that's stored in their muscles, the cache gathered during the time they spent at the surface.

No one has ever seen the battle between a sperm whale and a giant squid. But the stomachs of dead sperm whales contain collections of squid beaks, the only part of the squid that can't be digested. So each whale carries its own internal tally of fights won. As a successful whale swims back towards the sunlight, its lungs gradually re-inflate and reconnect with its blood supply. As the pressure decreases, the volume once again increases until it has reached its original starting point.

Oddly, the combination of complex molecular behaviour with statistics (not usually associated with simplicity) produces a relatively straightforward outcome in practice. There are indeed lots of molecules and lots of collisions and lots of different speeds, but the only two important factors are the range of speeds that the molecules are moving at, and the average number of times they collide with the walls of their container. The number of collisions, and the strength of each collision (due to the speed and mass of that molecule) determine the pressure. The push made by all that compared with the push from the outside determines the volume. And then the temperature has a slightly different effect.

*

'Who would normally be worried at this point?' Our teacher, Adam, is wearing a white tunic stretched over a happily solid belly, exactly what central casting demands of a jolly baker. The strong cockney accent is just a bonus. He pokes at the sad splat of dough on the table in front of him, and it clings on as though it's alive – which, of course, it is. 'What we need for good bread,' he announces, 'is air.' I'm at a bakery school being taught how to make focaccia, a traditional Italian bread. I'm pretty sure I haven't worn an apron since I was ten. And although I've baked lots of bread, I've never seen dough that looked like the splat, so I'm learning already.

Following Adam's instructions, we obediently start our own dough from scratch. Each of us mixes fresh yeast with water and then with the flour and salt, and works the dough with therapeutic vigour to develop the gluten, the protein that gives bread its elasticity. The whole time we're stretching and tearing the physical structure, the living yeast that's carried along in that structure is busy fermenting sugars and making carbon dioxide. This dough, just like all the others I've ever made, doesn't have any air in it at all – it just has lots of carbon dioxide bubbles. It's a stretchy sticky golden bioreactor, and the products of the life in it are trapped, so it rises. When this first stage is done, it gets a nice bath in olive oil and keeps rising, while we clean dough off our hands, the table, and a surprising amount of the surroundings. Each individual fermentation reaction produces two molecules of carbon dioxide which are expelled by the yeast. Carbon dioxide, or CO_2 – two oxygen atoms stuck to a carbon atom – is a small unreactive molecule, and at room temperature it has enough energy to float free as a gas. Once it's found its way into a bubble with lots of other CO_2 molecules, it will play bumper cars for hours. Each time it hits another molecule, there is likely to be some energy exchange, just like a cue ball hitting a snooker ball. Sometimes one will slow down almost completely and the other will take all that energy and zoom off at high speed. Sometimes the energy is shared between them. Every time a molecule bumps into the gluten-rich wall of the bubble, it pushes on the wall as it bounces off. At this stage, this is what makes the bubbles grow – as each one acquires more molecules on the inside, the push outward gets more and more insistent. So the bubble expands until the push back from the atmosphere balances the outward push of the CO_2 molecules. Sometimes the CO_2 molecules are travelling quickly when they hit the wall and sometimes they're travelling slowly. Bread bakers, like physicists, don't care which molecules hit which walls at particular speeds, because this is a game of statistics. At room temperature and atmospheric pressure, 29 per cent of them are travelling between 350 and 500 metres per second, and it doesn't matter which ones they are.

Adam claps his hands to get our attention, and uncovers the rising dough with a magician's flourish. And then he does something that is new to me. He stretches out the oil-covered dough and folds it over on itself, one fold from each side. The aim is to trap air between the folds. My initial unspoken response is 'That's cheating!', because I had always assumed that all the 'air' in bread was CO_2 from the yeast. I once saw an origami master in Japan enthusiastically teaching his students about the correct application of sellotape to an angular paper horse, and I felt the same unreasonable outrage then as in the bakery. But if you want air, why not use air? Once it's cooked, no one will know. I succumb to the knowledge of the expert and meekly fold my own dough. A couple of hours later, after more rising and folding and the incorporation of more olive oil than I had believed possible, my nascent focaccia and its bubbles were ready for the oven. The 'air' of both types was about to have its moment.

Inside the oven, heat energy flowed into the bread. The pressure in the oven was still the same as the pressure outside, but the temperature in the bread had suddenly gone up from 20°C to 250°C. In absolute units, that's from 293 Kelvin to 523 Kelvin, almost a doubling of temperature.[1] In a gas, that means that the molecules speed up. The bit that's counter-intuitive is that no individual molecule has its own temperature. A gas – a cluster of molecules – can have a temperature, but an individual molecule within it can't. Gas temperature is just a way of expressing how much movement energy the molecules have on average, but each individual molecule is constantly speeding up and slowing down, exchanging its energy with the others as they collide. Any individual molecule is just playing bumper cars with the energy it's got right now. The faster they travel, the harder they bump into the sides of the bubbles, so the greater the pressure they generate. As the bread went into the oven, gas molecules suddenly gained lots more heat energy and so they sped up. The average speed shifted from 480 metres per second to 660 metres per second. So the outward push on the bubble walls got much harder and the outsides weren't pushing back. Each bubble expanded in proportion to the temperature, pushing outward

[1] We'll get to the meaning of absolute temperature in chapter 6.

on the dough and forcing it to expand. And here's the thing . . . the air bubbles (mostly nitrogen and oxygen) expanded in exactly the same way as the CO_2 bubbles. This is the last piece of the puzzle. It turns out that it doesn't matter what the molecules are. If you double the temperature you still double the volume (if you keep the pressure constant). Or, if you keep the volume constant and double the temperature, the pressure will double. The complication of having a mix of different atoms present is irrelevant, because the statistics are the same for any mixture. No one looking at the final bread could ever tell which bubbles had been CO_2 and which ones had been air. And then the protein and carbohydrate matrix surrounding the bubbles cooked and solidified. The bubble size was fixed. Fluffy white focaccia was assured.

The way that gases behave is described by something called 'the ideal gas law', and the idealism is justified by the fact that it works. It works spectacularly well. It says that for a fixed mass of gas the pressure is inversely proportional to the volume (if you double the pressure, you halve the volume), the temperature is proportional to the pressure (if you double the temperature, you double the pressure) and the volume is proportional to the temperature, at fixed pressure. It doesn't matter what the gas is, only how many molecules of it there are. The ideal gas law is what drives the internal combustion engine, hot air balloons – and popcorn. And it applies not only when things heat up, but also when they cool down.

*

Reaching the South Pole was a major landmark in human history. The great polar explorers – Amundsen, Scott, Shackleton and others – are legendary figures, and the books about their achievements and failures are some of the greatest adventure stories of all time. And as if it wasn't enough to deal with unimaginable cold, lack of food, fierce oceans and clothing that wasn't up to the task, the mighty ideal gas law was against them, quite literally.

The centre of Antarctica is a high, dry plateau. It is covered in deep ice, but it hardly ever snows. The bright white surface reflects almost all of the feeble sunlight back into space, and temperatures can drop below −80°C. It is quiet. At an atomic level, the atmosphere here is sluggish, because the air molecules have little energy (due to the cold) and are moving relatively slowly. Air from above descends on to the plateau, and the ice steals its heat. Cold air becomes colder. The pressure is fixed, so this air shrinks in volume and becomes more dense. The molecules are closer together, moving more slowly, unable to push outwards hard enough to compete with the air around them pushing inwards. As the land slopes away from the centre of the continent towards the ocean, so this cold, dense air also slithers away from the centre along the surface, unstoppable, like a slow waterfall of air. It is funnelled through vast valleys, picking up speed as the funnels descend outwards, always outwards towards the ocean. These are the katabatic winds of Antarctica, and if you want to walk to the South Pole, they will be in your face all the way. It's hard to think of a worse trick nature could have played on those polar explorers.

'Katabatic' is just a name for this sort of wind, and it's found in many places, not all of them cold. As they descend, those sluggish molecules do warm up, just a little bit. And the consequences of that warming can be dramatic.

In 2007, I was living in San Diego and working at the Scripps Institution of Oceanography. As a northerner I was slightly suspicious of the eternal sunshine, but I got to swim in a 50-metre outdoor pool every morning so I couldn't really complain. And the sunsets were amazing. San Diego is on the coast with a clear view west across the Pacific Ocean, and the evening skyline was reliably stunning.

I really missed the seasons, though. It seemed as though time never moved on, almost like living in a dream. But then the Santa Ana winds came, and it went from sunny and warm and cheerful to sullenly hot and dry. The Santa Ana winds come every autumn, as air pours off the high deserts and flows over the coast of California out towards the ocean. These are also katabatic winds, just like the ones in Antarctica. But by the time they reach the ocean, the air is much hotter at the coast than it was on the high plateau. One memorable day, I was driving north up the I-5 freeway, towards one of the big valleys that funnelled the hot air out to sea. There was a river of low cloud sitting in the valley. My boyfriend at the time was driving. 'Can you smell smoke?' I asked. 'Don't be silly,' he said. But the next morning, I woke up in a weird world. There were huge wildfires to the north of San Diego, marching across the valleys, and there was ash in the air. A campfire had got out of control in the hot, dry conditions, and the winds were blowing the fire towards the coast. That river of cloud had been smoke. People went to work, and either were sent home or sat in huddles listening to the radio and wondering whether their houses were safe. We waited. The horizon was hazy because of ash clouds you could see from space, but the sunsets were spectacular. After three days, the smoke started to lift. People I knew had lost their houses to the flames. Everything had a layer of ash on it and health officials were advising against any outdoor exercise for a week.

Up on the high plateau, hot desert air had cooled, become more dense, and slithered downslope, just like the winds that faced Scott in Antarctica. But the wildfires started because that air wasn't only dry, it was hot. Why would it get hotter as it came downhill? Where does the energy come from? The ideal gas law still applies – this was a fixed mass of air, and it was moving so quickly that there was no time for it to exchange energy with its surroundings. As that stream of dense air made its way downhill, the atmosphere that was already at the bottom of the hill pushed on it, because the pressure down there was higher. Pushing on something is a way of giving it energy. You can imagine individual air molecules hitting the wall of a balloon that is moving towards them. They'll bounce off with more energy than they had to start with, because they're bouncing off a moving surface. So the volume of the air in the Santa Ana winds decreased because it was squeezed inwards by the surrounding atmosphere. That squeezing gave the travelling air molecules extra energy, and so the temperature of the wind increased. It's called adiabatic heating. Every year, when the Santa Ana winds come, everyone in California is extra vigilant about open fires. After a few days of such hot, dry air stealing the moisture from the landscape, sparks can easily turn into wildfires. And the heat doesn't just come from the California sun – it also comes from the extra energy given to the gas molecules as they are compressed by denser air closer to the ocean. Anything that changes the average speed of air molecules will change the temperature.

The same thing happens in reverse when you squirt whipped cream out of a can. The air that comes out in the cream has expanded suddenly and pushed on its surroundings, so it has given away energy and cooled down. The nozzle of the squirty cream canister feels cold to the touch for this reason – the gas that's coming through it is giving away its energy as it reaches the free atmosphere. Less energy is left behind, so the can feels cool.

Air pressure is just a measure of how hard all those tiny molecules are hammering on a surface. Normally we don't notice it much because the hammering is the same from every direction – if I hold up a piece of paper, it doesn't move because it's getting pushed equally from both sides. Each one of us is getting pushed by air all the time, and we hardly feel it at all. So it took people a long time to work out how hard that push actually is, and when it came along, the answer was a bit of a shock. The magnitude of the discovery was easy to appreciate because the demonstration was unusually memorable. It's not often that an important scientific experiment is also set up to be a theatrical spectacle, but this one had all the proper ingredients: horses, suspense, an astonishing outcome and the Holy Roman Emperor looking on.

The difficulty was that to work out how hard air is pushing on something, you really need to take away all the air on the other side of it, leaving behind a vacuum. In the fourth century BC, Aristotle had declared that 'nature abhors a vacuum', and that was still the prevailing view nearly a thousand years later. Creating a vacuum seemed out of the question. But some time around 1650, Otto von Guericke invented the first vacuum pump. Instead of writing a technical paper about it and disappearing into obscurity, he chose spectacle to make his point.[1] It probably helped that he was a well-known politician and diplomat, and was on good terms with the rulers of his day.

On 8 May 1654, Ferdinand III, the Holy Roman Emperor and overlord of a large part of Europe, joined his courtiers outside the Reichstag in Bavaria. Otto brought out a hollow sphere, 50 cm in diameter and made of thick copper. It was split into two separate halves with a smooth, flat surface where they touched. Each half had a loop attached to the outside, so that two ropes could be tied on and used to pull the halves apart. He greased the flat surfaces, pushed the two sides together, and used his new vacuum pump to remove the air from the inside of the sphere. There was nothing to hold it together, but after the air had been removed, the two halves behaved as though they were glued to each other. Otto had realized that the vacuum pump gave him a way to see how strongly the atmosphere could push. Billions of minuscule air molecules were hammering all over the outside of the sphere, pushing the halves together. But there was nothing

---

[1] This substitution is not the recommended way of doing science today.

inside to push back.[1] You could only pull the two hemispheres apart if you could pull harder than the air could push.

Then the horses were assembled. A team was hitched to each side of the sphere, pulling in opposite directions in a giant tug of war. As the Emperor and his retinue looked on, the horses strained against the invisible air. The only thing holding the sphere together was the force of air molecules hitting something the size of a large beach ball. But the strength of thirty horses could not pull the sphere apart. When the tug of war had finished, Otto opened the valve to let air into the sphere, and the two halves just fell open. There was no question about the winner. Air pressure was far stronger than anyone had suspected. If you take all the air out of a sphere that size and hang it vertically, the upward push of the air could theoretically support 2,000 kg, the weight of a large adult rhino. That means that if you draw a circle 50 cm in diameter on the floor, the push of the air on just that bit of floor is also equal to the weight of a 2,000 kg rhino. Those tiny invisible molecules are hitting us very hard indeed. Otto did this demonstration many times for different audiences, and the sphere became known as a Magdeburg sphere, named after his home town.

Otto's experiments became famous partly because others wrote about them. His ideas first reached the scientific mainstream in a book by Gaspar Schott, published in 1657. It was reading about Otto's vacuum pump that inspired Robert Boyle and Robert Hooke to carry out their experiments on gas pressure.

You can try a version of this for yourself, without the need for either horses or emperors. Find a square of thick, flat cardboard that's large enough to cover the mouth of a glass. It's best to try this over a sink, just in case. Fill up the glass with water right to the rim and put the cardboard on top. Push it flat against the rim

---

[1] We don't know how much of the air Otto's vacuum pump removed. It won't have been all the air, but it must have been a substantial proportion of it.

of the glass so there's no air left between the surface of the water and the cardboard. Then turn the glass over – and remove your hand. The cardboard, supporting the entire weight of the water, will stay put. It stays there because air molecules are hitting it from the underside, pushing the cardboard upwards. That push is easily enough to hold the water up.

The battering of air molecules isn't just useful for keeping things in place. It can also be used to move things around, and humans weren't the first ones to take advantage of that. Let's meet an elephant, one of the Earth's most impressive experts at manipulating its environment with air.

An African bush elephant is a majestic giant, usually found ambling peacefully across dusty dry savannah. Elephant family life is based around groups of females. An elder stateswoman, the matriarch, leads each group as they roam in search of food and water, relying on her memory of the landscape to make decisions. But these animals don't just depend on their heft to survive. Each elephant may have a heavy lumbering body, but to make up for it, it's got one of the most delicate and sensitive tools in the animal kingdom: a trunk. As a family group moves around, they're constantly exploring the world with this odd appendage, signalling, sniffing, eating and snorting.

An elephant's trunk is fascinating in many ways. It's a network of interlocking muscles, capable of bending and lifting and picking up objects with incredible dexterity. If that were all it is, it would be useful enough, but it's made even better by the two nostrils that run down the length of the trunk. These nostrils are flexible pipes that join the snuffling trunk tip to the elephant's lungs, and this is where the real fun starts.

As our elephant and her family group approach a watering hole, the 'still' air around them is bumping and jostling just like anywhere else, hammering against their wrinkly grey skin, against the ground and against the water surface. The matriarch is slightly ahead of the group, swinging her trunk as she saunters into the pool and sends ripples through her reflection. She dips her trunk into the water, closes her mouth, and the huge muscles around her chest lift and expand her ribcage. As her lungs expand, the air molecules inside spread out to take up the new space. But that means that right down inside the tip of the trunk, where the cool water touches the air in her nostrils, there are fewer air molecules hitting the water. The ones that are there are going just as fast, but there aren't as many collisions. The consequence is that the pressure inside her lungs has dropped. Now the atmosphere is winning in the shoving contest between the air molecules hitting the pool of water and the air molecules inside the matriarch. The push from inside can't match the push from outside any more, and the water is just the stuff in the middle of the competition. So the atmosphere pushes water up the elephant's trunk, because what's inside can't push back. Once the water has taken up some of the extra space, the air molecules inside are as close together as they were at the start, and the water doesn't move any further.

Elephants can't drink through their trunks – if they tried, they'd cough just as you would if you tried to drink through your nose. So once the matriarch has perhaps 8 litres of water inside her trunk, she stops expanding her ribcage. Curling her trunk up and under, she points the tip into her mouth. Then she uses her chest muscles to squash her chest, decreasing the size of her lungs. As the air molecules inside are squashed closer together, the water surface halfway up her trunk gets hit much more often. The battle of the air inside and the air outside reverses, and the water is pushed out of the trunk into the elephant's mouth. Our matriarch is controlling the volume of her lungs to control how hard the air inside her is pushing on the outside. If she shuts her mouth, the only place where anything can move is at her trunk, and whatever is at the tip of her trunk will get pushed in or pushed out. An elephant's trunk and lungs together are a combined tool for manipulating air so that the air, rather than the elephant herself, does the pushing.

We do the same thing when we suck a drink up a straw.[1] As we expand our lungs, the air inside is spread more thinly. There are fewer air molecules inside the straw to push on the water surface. And so the atmosphere pushing on the rest of the drink pushes the drink up the straw. We call this sucking, but we're not pulling on the drink. The atmosphere is pushing it up the straw, doing the work for us. Even something as heavy as water can be shunted about if the hammering of the air molecules is harder on one side than the other.

However, sucking air up a trunk or a straw has limits. The bigger the pressure difference between the two ends, the harder the push will be. But the biggest difference you can possibly make when you're sucking is the difference between the pressure of the atmosphere and zero. Even with a perfect vacuum pump instead of lungs, you couldn't drink through a vertical straw that was longer than 10.2 metres, because our atmosphere can't push water any higher than that. So to exploit to the full the ability of gas molecules to push things around, you need to get them working at higher pressures. The atmosphere can push hard, but if you force another gas to be hotter and put it under greater pressure, it can push harder. Get enough tiny gas molecules hitting something often enough and fast enough, and you can move a civilization.

A steam locomotive is a dragon made of iron, a hissing, breathing, muscular beast. Less than a century ago these dragons were everywhere, hauling the products of industry and the needs of society across whole countries and expanding the world of their passengers. They were mundane and noisy and polluting, but they were beautiful pieces of engineering. When they became obsolete, the dragons were not allowed to die because society just couldn't let them go. They've been kept alive by volunteers,

[1] And also when we breathe. Every breath you take gets into your lungs because the atmosphere pushes it there.

enthusiasts and a deep well of affection. I grew up in the north of England, and so my childhood years were steeped in the history of the Industrial Revolution: mills, canals, factories and, more than anything else, steam. But I live in London now, and so it's easy to forget. A trip along the Bluebell steam railway with my sister brought it all back.

It was a chilly winter day, absolutely perfect for a journey propelled by steam with the promise of tea and scones at the other end. We didn't spend too long at the station where we started, but when we arrived at Sheffield Park, we stepped off the train into a slow but steady hum of activity. The engines were constantly being tended by an ever-changing swarm of humans who seemed tiny beside their iron beasts. The humans involved with the engines were easy to identify: blue overalls, peaked caps, a jolly demeanour, an optional beard and in between engine-tending duties usually to be found leaning on something. As my sister pointed out, an awful lot of them seemed to be called Dave. The beauty of a steam engine is that the principle behind it is fantastically simple, but the raw power produced needs to be goaded, tamed and nurtured. A steam engine and its humans are a team.

Standing on the ground, looking up at one large, black engine, it was hard to comprehend that in its heart, this was basically a furnace on wheels heating a giant kettle. One of the Daves invited us into the cab. We climbed up the ladder just behind the engine and found ourselves in a grotto full of brass levers, dials and pipes. There were also two white enamel mugs and a sandwich tucked behind one of the pipes. But the best thing about the cab was that we could see right into the belly of the beast. The giant furnace at the heart of a steam engine is filled with fiery coals burning an intense yellow. The fireman gave me a shovel and told me to feed it, and so I obediently scooped coal from the tender behind me into the glowing mouth. The engine is hungry. On one 18-km journey, it will burn through 500 kg of coal. That half-tonne of solid black gold is converted into gas – carbon dioxide and water; and the burning releases enormous amounts of energy, so those gases are extremely hot. This is the start of the energy conversion that powers the train.

When you look at a steam engine, the main feature is the long cylinder of the 'engine' itself, stretching from the cabin to the funnel. I'd never really thought about what was in there, but it's full of tubes. The tubes are carrying the hot gas from the firebox through the engine, and this is the kettle. Most of the space around the tubes is taken up by water, a giant bath of bubbling, boiling liquid. As this is heated by the tubes, it produces steam, hot water molecules that are zooming about in the space right at the top of the engine at very high speeds. This is what most of the steam engine is: furnace and kettle, producing vast clouds of hot water vapour. This dragon isn't breathing fire, it's breathing billions of energetic molecules, all whizzing about at gigantic speeds but trapped in the engine. The temperature of that gas is about 180°C and the pressure in the top of the kettle is about ten times as high as atmospheric pressure. The molecules are thumping hard against the walls of the engine, but they can only escape after being put to work.

We climbed down from the cab and walked up to the front. The towering engine, the half-tonne of coal, the giant kettle and the human teamwork were all in service to what we found there: two cylinders containing pistons, each about 50 cm in diameter and 70 cm long. It's down here at the front, dwarfed by the dragon above, that the real work is done. The hot, high-pressure steam is fed into one cylinder at a time. The atmospheric pressure on the other side of the piston is no match for the ten atmospheres that the dragon has breathed out. The hammering molecules shove the piston along the cylinder, and then are finally released to the atmosphere with a satisfied 'chuff'. This is what you're hearing when a steam engine's familiar 'chuff, chuff, chuff' comes towards you. It's the release to the atmosphere of water vapour whose work is done. The piston drives the wheels, and the wheels grip the rails and drag the carriages. We know that steam engines need vast quantities of coal to keep them going, but almost no one talks about the water used on every journey. The 500 kg of coal that is shovelled into this engine on each trip is used to convert 4,500 litres of water to gas, and then that gas pushes on a piston and is lost to the atmosphere one 'chuff' at a time.[1]

Finally it was time to leave the engine and get back in one of the carriages to be carried home. The return journey felt different. The billows of steam whooshing past the windows had made their contribution to our excursion. Instead of appearing loud and intrusive, the engine pulling us along seemed relatively quiet and calm considering what was going on inside it. It would be lovely if someone could make a glass steam locomotive one day, so that we could all see the beast at work.

The steam revolution in the early 1800s was all about using the push of gas molecules to do something useful. All you need is a surface with gas molecules hitting one side harder than on the other. That push could lift the lid of a pan as you cook, or it could be used to transport food and fuel and people, but it comes from the same basic principles. We don't use steam engines any more, but we do still use that push. A steam engine is technically an 'external combustion engine' because the furnace is separate from the kettle. In a car engine, the burning happens in the cylinder – gasoline burns right next to the piston and the burning itself produces hot gas to shove the piston along. That's classed as an internal combustion engine. Every time you get

---

[1] If you've ever wondered what Thomas the Tank Engine's tank is all about, it's all about the water. The water can be stored in a separate carriage with the coal (a tender) or it can be stored in a tank that sits around the engine. Thomas stores his water around the engine – that's why he was rectangular – and so he's a tank engine.

into a car or a bus, you're being carried along by the push of gas molecules.

It's easy to play with the effects of pressure and volume, especially if you can find a wide-necked bottle and a hard-boiled egg with its shell taken off. The neck of the bottle needs to be just a bit narrower than the egg, so that the egg will perch on top of it without falling in. Light some paper, drop it into the bottle, let it burn for a few seconds and then put the egg back on top. After a while, you'll see the egg squeeze itself down inside the bottle. That's a bit weird, and it's inconvenient that now you have an egg in a bottle and it won't come out. There are a few solutions, but one of them is to turn the bottle upside-down so that the egg is sitting in the neck, and then run the bottle under a hot tap. After a while, the egg will come whooshing out.

The game here is that you have a fixed mass of gas (in the bottle) and a way to tell whether the pressure inside is higher or lower than the pressure of the atmosphere. If the egg is blocking the neck, the volume of gas inside is fixed. If you increase the temperature by setting fire to something, the pressure inside will increase and air will escape around the sides of the egg (if the egg is sitting on top). When it cools down again, the pressure inside will decrease (since the volume is fixed) and the egg will be pushed inside, because the push from outside is now greater than the push back from inside. You can get the egg to move just using the heating and cooling of air in a container with a fixed volume.

The high pressures in a steam engine are controlled and stable, ideal for pushing pistons and making wheels turn. But that's not the end of it. Why waste energy on intermediate stages between the gas and the wheels? Why not just let the hot high-pressure gases shove your vehicle forward directly? That's how guns, cannons and fireworks have always worked, although the early ones were all notoriously unreliable. But by the early 1900s, technology and ambition had moved on. Along came the rocket, the most extreme form of direct propulsion ever invented.

It wasn't until after the First World War that the necessary technology reached any degree of reliability, but by the 1930s you could launch a rocket that would probably go in the right direction and probably wouldn't kill anyone. Most of the time. As with many new technologies, inventors made it work before anyone knew what to do with it. And out of the fertile pond of enthusiastic human inventiveness came something very new and modern-sounding and utterly doomed: rocket post.

In Europe, rocket post only really happened because of one man: Gerhard Zucker. A few inventors at that time were tinkering with rockets, but Zucker led the field in dogged persistence and unfailing optimism in the face of continual discouragement. This young German was obsessed by rockets, and since the military weren't interested in what he was doing he looked to the civilian world for an excuse to continue. Sending mail by rocket sounded to him like something the world was crying out for – fast, capable of crossing the sea and covered in the glitter of novelty. The Germans tolerated his early (unsuccessful) experiments and then decided they'd had enough, so Zucker came to the UK. Here he found friends and support in the stamp-collecting community, who liked the idea of a new kind of novelty stamp to go with a new kind of novelty mail delivery system. Things were looking up. After a small-scale test in Hampshire, Zucker was sent up to Scotland in July 1934 to test sending his mail rocket between two islands, Scarp and Harris.

Zucker's rocket wasn't particularly sophisticated. The main body of it was a large metal cylinder about a metre long. Inside, a narrow copper tube with a nozzle at its back end was filled with packed powder explosive. The space in between the inner tube and outer cylinder was filled with letters, and there was a pointy nose on the front with a spring in it, presumably to help soften the landing. Rather sweetly, on Zucker's diagram of the setup, the thin layer between the explosive and the highly flammable letters is labelled 'asbestos packing round cartridge, to prevent damage to mails'. The rocket was laid down on its side on a slanted trestle, pointed upwards and sideways. At the moment of launch, a battery would ignite the explosive, and the burning would produce vast quantities of hot, high-pressure gas. The gas molecules, now moving at high speed, would bounce off the inside of the front end of the rocket, driving it forward, but there would be no equivalent push at the back end – gas would just escape through the nozzle to the atmosphere. This imbalance in pushing could drive the rocket forward very quickly. The explosive burn would continue for a few seconds, enough to push the rocket high into the air and over the channel between the islands. There didn't seem to be too much concern about how and where it would land, but that was one reason for trying it out in a very remote part of Scotland surrounded by sea.

Zucker collected 1,200 letters to send as part of the trial, each adorned with a special stamp that said 'Western Isles Rocket Post'. He packed as many as would fit inside his rocket and set up the trestle, watched by a bemused crowd of locals and an early BBC TV camera. The moment had come.

When the launch button was pressed, the battery ignited the explosive. The rapid burning generated the expected mixture of hot gases inside the copper tube, and the energetic molecules hammered on the front of the rocket, shoving it up the trestle at high speed. But after only a couple of seconds, there was a loud, dull thud and the rocket disappeared behind a plume of smoke. As the smoke cleared, hundreds of letters could be seen fluttering to the ground. The asbestos had done its job, but the rocket hadn't. Hot, high-pressure gas is hard to control, and the energetic molecules had broken the casing. Zucker blamed the explosive cartridge, and set about collecting the letters and preparing for a second trial.

A few days later, 793 surviving letters from the first rocket and also 142 new ones were packed into a second rocket. This one was launched from the other island, Harris, back towards

Scarp. But Zucker was out of luck. The second rocket also exploded on the launch pad, this time with an even louder bang. The surviving letters were collected up again and sent to their recipients by the conventional mail system, with singed edges as souvenirs. The trial was abandoned. For the next few years, Zucker stubbornly carried on, always convinced that next time, it would work. But it never did,[1] at least not for mail. Zucker pushed hard against the unknown, and it's only hindsight that tells us that it wasn't the right time or the right place or the right idea. If it had been all three, we'd hail him as a genius. But small-scale rocketry was just too unreliable and fiddly to deliver messages better and faster than motorized transport and the telegraph. In a way he was right: using hot, high-pressure gases as a propellant has enormous potential to get things from A to B. But it was others who took the principle, found a suitable application, and solved the practical problems until it became a success. Rocket development became the preserve of the military, with the German V1 and V2 rockets used in the Second World War showing the way, and civilian space programmes taking over after that.

These days, we are all familiar with images of giant rockets carrying huge cargoes of people and equipment to the International Space Station, or taking satellites into orbit. Rockets can seem frighteningly powerful, and the modern control systems that now make them safe and reliable are a huge human achievement. But the basic mechanism behind every Saturn V rocket, every Soyuz and Arianne and Falcon 9 that has ever flown, is the same as it was for Gerhard Zucker's primitive mail rocket. If

[1] The Indian Airmail Society also experimented with rocket mail around the same time. They managed 270 flights, sending parcels as well as letters, but never established it as a long-term success. In the end, rocket mail was never going to be able to compete with the regular ground-based mail delivery systems on reliability and cost.

you make enough hot, high-pressure gas quickly enough, you can make use of the huge cumulative force that comes from billions of individual molecules bumping into things. The flight pressure in the first stage of a Soyuz rocket is about sixty times greater than atmospheric pressure, so the push is sixty times stronger than the normal push of the air. But it's exactly the same type of push: just molecules bumping into things. Vast quantities of them colliding often enough and fast enough can send a man to the Moon. Never underestimate things that are too small to see!

Gas molecules are always with us. The Earth has an atmosphere that surrounds us, bumps into us, pushes on us and also keeps us alive. The wonderful thing about our atmosphere is that it isn't static – it's constantly shifting around and changing. The air is invisible to us, but if we could see it, we'd see huge masses of it heating up and cooling down, expanding and contracting, always moving. What our atmosphere is doing is dictated by the gas laws we've seen at work in this chapter, just like any other collection of gas molecules. Even though it's not contained in a whale's lungs or a steam engine, it's still pushing. But since its surroundings are also air, that means that it's constantly pushing itself around, readjusting to conditions. We can't see the details, but we have a name for the consequences: weather.

The best place to watch a storm is a vast open plain. The day before, the air can be calm and the expanse of blue up above seems to go on for ever. Invisible air molecules crowd together close to the ground and spread out further up, always pushing, hassling, readjusting and flowing. Air is shunted from regions of high pressure to regions of low pressure, responding to heating and cooling, always on the way to somewhere else. But the adjustments are slow and peaceful, and there's no hint of the vast amounts of energy carried by the molecules.

The day of the storm dawns just as the one before it did, but the sky is clearer, so the ground is heating up more quickly. The air molecules take some of that energy and speed up. By early afternoon, a deep wall of cloud is approaching and expanding as it moves, until it stretches across the horizon. Energy is on the move. A pressure difference is pushing this slab of gaseous architecture across the plain. The drama comes because this giant structure is unstable. Although the air molecules are shoving hard on each other, they haven't had time to rearrange themselves into a more balanced situation. Alongside that, vast amounts of energy are being shunted around, so the situation is constantly changing. Hot air warmed by the ground is pushing upwards into the cloud, pummelling its way through and building towers that stretch high above the wall.

As the thundercloud arrives overhead, the expansive blue is replaced by a dark low lid on the landscape. On the ground, we are boxed in by the clash that is going on above. We can't see the air molecules, but we can see the clouds churning and surging. And this is only a hint of the violence going on within them as air packets are buffeted and pummelled, because the imbalances of pressure are so strong that readjustment is a rapid and energetic process. As energy is exchanged by the air molecules, water droplets cool and grow and the first large raindrops start to fall. Strong winds stream past us, as the air molecules rush around even at ground level.

Big storm clouds remind us how much energy there is up there in the blue sky. We see hints of the bumping and shoving, and it looks extreme – but it's only the merest hint of the real bumping and shoving happening at a molecular level above our heads. Air molecules may absorb energy from the Sun, lose energy to the ocean, gain energy from condensation as clouds form or lose energy by radiating it away to space, and they are constantly adjusting according to the ideal gas law. Our spinning planet with its rough and multicoloured surface makes the adjustments more complicated, and so do clouds, tiny particulates, and the specific gases present. A weather forecast is really just a way of keeping track of the battles above our heads and picking out the ones that will affect us most down here on the ground. But the action right at the root of it all is the same as that used by an elephant, a rocket and a steam engine. It's all just the gas laws in action. The same bit of physics that makes popcorn pop also makes the weather work.

2

What goes up must come down

Gravity

CURIOSITY RUNS IN MY family. They'll happily go and investigate anything new that comes along, they're up for trying things out, and they do all this without making any fuss about it. So they weren't all that surprised when I disappeared off to the kitchen during a family lunch on a suddenly urgent mission to find a bottle of lemonade and a handful of raisins. It was a beautiful summer day, and we were all sitting outside in my mother's garden: my sister, aunt, Nana and my parents. I found one of those 2-litre plastic bottles of cheap fizzy lemonade, took the label off, and then put the bottle on the middle of the table. This new madness was watched with quiet interest, but I had their attention, so I took the cap off and dropped the entire handful of raisins into the bottle. There was a whoosh of foam and then, when the bubbles cleared, we could see that the raisins were dancing. I had thought that this would only be entertaining for a minute or two, but Nana and my Dad couldn't stop staring at it. The bottle had been transformed into a raisin lava lamp. The raisins were rushing from the bottom of the bottle to the top and back again, twirling madly and bumping into each other on the way.

A sparrow landed on the table to hoover up some crumbs and gave the bottle a suspicious look. My Dad was giving the bottle a suspicious look from the other direction. 'Does it only work with raisins?' he asked.

The answer is yes, and for a really good reason. Before you

take the lid off a fizzy drink, the pressure inside is significantly higher than the pressure of the air around you. At the moment you unscrew the cap, the pressure inside the bottle drops. There's lots of gas dissolved in the water, kept there by the higher pressure, but suddenly all that gas can escape. The problem is, it needs a route out. Starting a new gas bubble is really difficult, so the gas molecules can only join an existing bubble. What they need is a raisin. Raisins are helpfully covered in V-shaped wrinkles that won't have been completely filled by the lemonade. Down at the bottom of each wrinkle there's a proto-bubble, a tiny pocket of gas. This is why you need raisins, or something else that's small, wrinkly and just a tiny bit more dense than water. Gas floods out of the lemonade into those proto-bubbles, and each raisin grows itself a bubbly lifejacket that stays stuck to the raisin. By themselves, raisins are more dense than water, so gravity drags them down to the bottom. But after they've grown a few bubbles they become less dense overall, and they begin their journey up to the top. Once they get there, the bubbles that break the surface pop, and you can see the raisins tip over as the bubbles underneath lift themselves up and pop in turn. Once there is no lifejacket left, the raisin is more dense than the lemonade, so down to the bottom it sinks. This will keep going until all the excess carbon dioxide gas has come out of the lemonade.

After half an hour as the table's centrepiece, the frantic dance of the raisins had been reduced to an occasional leisurely excursion to the surface, and the lemonade had turned an off-putting yellowish colour. Beautiful buoyant exuberance had been transformed into what looked like a giant urine sample bottle with dead flies at the bottom.

Try it. It's a good way of cheering up a slightly dull party if you can find raisins or currants in the party snacks. The key is that the bubbles and the raisin become a single object and move about as one. If you puff the raisin up with portable air pockets, you barely make it any heavier, but the whole thing takes up a lot more space. The ratio of 'stuff' to space filled is density, so the raisin-plus-bubble combination is less dense than the raisin alone. Gravity can only pull on 'stuff', so things that are less dense feel less of a pull towards the Earth. This is why some objects can float – floating is just a gravitational hierarchy sorting itself out. Gravity pulls dense liquids downwards, and any less dense object in the liquid is relegated to floating about on top. We say that anything that's less dense than a liquid is buoyant.

Air-filled spaces are really useful for controlling relative density, and therefore buoyancy. Famously, one of the design features that was supposed to make the *Titanic* 'unsinkable' was the large watertight compartments that took up the lower part of the ship. They were acting like the bubbles on the raisin: air-filled pockets that made the ship more buoyant and kept it afloat. When the *Titanic* got into trouble, those compartments turned out not to be watertight, and as they filled with water, the effect was the same as the last few bubbles popping at the surface. Just like the raisin without its lifejacket, the *Titanic* had to sink down to the depths.[1]

We accept that things sink and float, but rarely think about the real cause: gravity. The theatre of our lives is played out on a stage dominated by this one ever-present force, always making it very clear which way is 'down'. It's fantastically useful – it keeps everything organized by keeping it on the floor, for a start. But it's also the most obvious single force to play with. Forces are weird – you can't see them and it can be hard to know what they're up to. But gravity is always there, with the same strength (at the Earth's surface, at least), and pointing in the same

---

[1] By coincidence, the distance that the *Titanic* sank relative to its size (14 times its length) is pretty much the same as the distance that the raisins sink in a 2-litre bottle (a large raisin is about 2 cm long, and the bottle is about 30 cm deep). The *Titanic* was 269 metres long, and sank in water that was 3,784 metres deep.

direction. If you want to play with forces, gravity is a great place to begin. And how better to begin playing than by falling?

Springboard and highboard diving sit somewhere on the scale between utter freedom and complete madness. The moment you leave the board, you are completely free of the feeling of gravity. It's not that it's gone away, it's just that you're giving in to it completely, so there's nothing left to push against. You can rotate just like a theoretical free body, as if you were floating in space, and it's incredibly liberating. But there is no such thing as a free lunch, and the problem comes a second or so later, when you arrive at the water's surface. There are two ways of dealing with it. You can either make a small tunnel into the water with your hands or feet, and organize yourself so that the rest of your body slides gracefully into the tunnel, minimizing the splash. Or you can let your arms and legs, tummy or back each make their own impact, generating a very large splash. That second one hurts.

I was a springboard diver and coach for a few years in my twenties, but I hated highboard diving. The springboards are the bouncy ones, 1 and 3 metres above the pool. It's a bit like trampolining, but with a softer landing. Highboards are the solid platforms above that, at 5, 7.5 and 10 metres. The pool I trained at only had a 5-metre platform, but I did everything I could to avoid it.

Up on the 5-metre platform, the water looks a long way away. There was always a thin stream of bubbles coming from below, so that you could see where the water surface was even if the pool was completely still. The most basic warm-up dive is a 'front fall': exactly what it says on the tin. Standing on the end of the board, you lean forward into an 'L' shape with your arms locked above your head, keeping your body straight apart from the bend at the hips. Things look a tiny bit less terrifying from here, because your head is closer to the water, but not much. Then you lift yourself up on tiptoe, and surrender. Suddenly, you are free. There is just you and a planet with a mass of

6 million billion billion kg, linked only by this thing called gravity, and the laws of the universe mean that you are pulling on each other.

Gravity, like any other force, changes your speed – it accelerates you. This is a consequence of Newton's famous second law,[1] which states that any net force acting on you will change your speed. When jumping off a diving board, you're stationary to begin with, so you start to move slowly. The interesting thing about acceleration is that it's measured in units of change of speed per second. At the start, you're only getting going, so it takes a relatively long time (0.45 seconds) to fall through the first metre. But you go through the second metre much more quickly, so there's less time to accelerate you on the way. After one metre, your speed is 4.2 metres per second, but after two metres it's only 6.2 metres per second.

So you spend most of your time during a dive in the worst place, high above the water. In the first half of the time you spend in the air on a 5-metre dive, you only fall 1.22 metres. After that, things happen very quickly. Falling the full 5 metres takes 1 second, and by the end of that fall you're travelling at 9.9 metres per second. You straighten out your body, reach for the water and hope for a splash-free entry.

When competitions came around, the others in the group would eagerly take the opportunity to compete from the higher boards at whichever pool we were visiting. I did not. As far as I was concerned, more time in the air meant more time for things to go wrong. But this was never particularly logical, because you're travelling so fast that falling through the extra distance doesn't actually speed you up that much. It takes 1 second to fall 5 metres, but only 1.4 seconds to fall 10 metres. And you're only going 40 per cent faster, even though you've fallen twice as far. I knew that. But I was a diver for about four years and I have never

---

[1] It's often written: Force = mass × acceleration, or $F = ma$.

once jumped off a board higher than 5 metres. I'm not scared of heights. I'm just scared of impacts. The longer gravity has had to accelerate me, the less pleasant it's likely to be during the deceleration phase. Even dropping your phone is a reminder that letting gravity take over isn't always a good idea. Extra distance to fall still provides the opportunity for extra speed . . . except when it doesn't.

On Earth, there is a limit to what gravity can do to you. That's because you are only accelerated by the overall force on you, called the resultant force. As you speed up, you have to push more air out of the way in a given period of time, and that air also pushes back on you, effectively reducing the pull of gravity because it's pushing in the opposite direction. At some point, those two things balance, and you will travel at your terminal velocity, unable to get any faster. For leaves and balloons and parachutes, the force of the air pushing back is pretty big compared with the weedy gravitational pull, and so that force balance is reached at a relatively low speed. But for a human, terminal velocity close to the ground is around 120 mph. Sadly for any falling humans, air resistance is pretty negligible until you get to very high speeds. And it certainly doesn't push back enough to reassure me about jumping off a 10-metre highboard, even now.

*

My scientific research is all about the physics of the ocean surface. I'm an experimentalist, and so part of my job is to go out on the ocean and measure what's happening at this messy, beautiful boundary between the air and the sea. And that means spending weeks working on a research ship, a floating, functional, mobile scientific village. The problem with living on a ship is that you have to live with gravity basically having gone wrong. 'Down' becomes an uncertain concept. Things may fall at the same speed and in the same direction as if you'd dropped them on land, but then again, they may not. If you spot a loose object just sitting on a table, you tend to find yourself watching it suspiciously because there is no guarantee that it's going to stay put. Life at sea is full of elastic bungees, string, rope, sticky grippy mats, locked drawers – anything that helps to keep life organized when there's a capricious force pulling things in unpredictable directions, like a scientific poltergeist. My specific research topic is the bubbles produced by breaking waves in storms, and so I've spent months living at sea in some pretty nasty conditions. I actually quite like it – you adapt very quickly – but it's a good lesson in how much we take gravity for granted. On one research ship in the Antarctic, the ship's purser used to marshal the unreasonably enthusiastic among us for circuit training three times each week. We'd gather in the hold, an echoing iron space down in the guts of the ship, and obediently jump and lift and skip for an hour. It was probably the most effective circuit training I've ever done, because you never knew what force you were going to have to resist. The first three sit-ups might be ridiculously easy, because the ship was heaving downwards, effectively reducing gravity. You'd just start feeling really good about yourself when the penalty arrived as the ship reached the bottom of the trough of the wave. At that point, gravity was effectively 50 per cent stronger, and suddenly it felt as though your tummy muscles had to fight against strips of elastic pinning you to the floor. Four more sit-ups and gravity would vanish again . . . Anything involving jumping was even worse because you were never quite sure where the floor was. And then afterwards, in the shower, you'd spend your time chasing the water flow around the shower cubicle, as the rolling of the ship made it impossible to predict where it was going to fall.

Of course, there was nothing wrong with gravity itself. Everything on that ship was being pulled towards the centre of the Earth with the same force. But when you feel the force of gravity, you're resisting an acceleration. If your surroundings are accelerating all by themselves as the giant tin can you're living in is tossed around by nature, your body can't tell the difference between gravitational acceleration and any other acceleration that's going on. So you get 'effective gravity', which is what you experience overall, without worrying about where it's coming from. That's why that odd feeling you get in a lift only happens at the beginning and end of the ride – when the lift is accelerating towards its top speed, or decelerating (a negative acceleration) towards a halt. Your body can't tell the difference between the acceleration of the lift and the acceleration due to gravity,[1] so you experience a reduced or increased 'effective gravity'. For a fraction of a second, you can experience what it might be like to live on a planet with a different gravity field.

Fortunately for us, we're completely free of these complications most of the time. Gravity is constant and points towards the centre of the Earth. 'Down' is the direction in which things fall. Even plants know that.

My mother is a keen gardener, so when I was growing up I had a lot of opportunities to plant seeds, chop up weeds, wrinkle my nose in disgust at slugs and turn over compost heaps. I remember being fascinated by seedlings because they clearly knew up from down. Down in the darkness of the soil, a seed case would open and new roots would creep downwards while a nascent shoot explored upwards. You could pull up a young seedling and see that there had been no hesitation or exploration. The root just went straight down and the shoot went

---

[1] If you've ever wondered what General Relativity is really about, the core of it is just this realization. If you're in a closed lift, whether you're standing, playing catch or doing sit-ups, you can't tell which forces are due to 'gravity' and which are because the lift is accelerating. Einstein realized that there is a way of looking at what matter does to space which shows that these forces are indistinguishable because they're actually exactly the same thing.

straight up. How did it know? When I was a bit older, I found the answer and it's delightfully simple. It turns out that inside the seed there are specialized cells called statocytes that are mini plant snow-globes. Inside each one there are specialized starch grains that are more dense than the rest of the cell, and they settle towards the bottom of the cells. Protein networks can sense where they are, and so the seed, and later the plant, knows which way is up. Next time you plant a seed, turn it over and think about the mini snow-globe inside, and then plant it which-ever way up you like, because the plant can solve the puzzle.

Gravity is a fantastically useful tool. Plumb lines and spirit levels are cheap and accurate. 'Down' is universally accessible. But if everything is pulling on everything else, what about the mountain that I can see in the distance? Isn't that pulling on me? What's so special about the centre of the planet?

I love coastlines for all sorts of reasons (waves, bubbles, sun-sets and sea breezes), but what I love most of all is the liberating, luxurious feeling of taking in the vast expanse of the sea. When I lived in California, I shared a tiny house very close to the beach, so close that we could hear the waves at night. There was an orange tree in the back garden, and a porch for watching the world go by. The ultimate luxury at the end of a busy day was to walk to the end of the road, sit on the smooth, worn rocks and look out at the Pacific Ocean. When I did that sort of thing in England as a kid, I was just watching for fish or birds or big waves. But when I watched the ocean in San Diego, I was imag-ining the planet. The Pacific Ocean is vast, taking up a full third of the circumference of the Earth at the equator. Looking out towards the sunset, I could imagine the giant ball of rock I was living on, Alaska and the Arctic far away to my right, in the north, and the full length of the Andes running all the way to Antarctica on my left, south of me. I could almost give myself vertigo watching it all in my head. And once, it occurred to me that I was directly experiencing all of those places. Each one of them was tugging on me, and I was tugging on them. Every bit of mass is pulling on every other bit of mass. Gravity is a fantastically weak force – even a small child can generate the force to resist the gravitational pull of a whole planet. But nevertheless, each of those minuscule tugs is still there. Together, an uncountable number of minute tugs adds up to a single force, the gravity that we experience.

This was the step taken by the great scientist Isaac Newton when he published his Law of Universal Gravitation in *Philosophiae Naturalis Principia Mathematica* – the famous *Principia* – in 1687. Using the rule that the gravitational force between two things is inversely proportional to the square of the distance separating them, he showed that if you added up the pull of every single bit of a planet, quite a lot of those sideways pulls cancelled each other out, and the result was a single downward force, pointing towards the centre of the planet and proportional to the Earth's mass and the mass of the thing being pulled. A mountain that's twice as far away will only pull on you with a quarter of the force. So distant objects matter less. But they still count. Sitting looking out at the sunset, I was being pulled sideways to the north and a bit downwards by Alaska and sideways to the south plus a bit downwards by the Andes. But the pulls to the north and the south cancelled each other out, and what was left over was downwards.

So even though we're all being pulled on (right now) by the Himalaya, the Sydney Opera House, the Earth's inner core and lots of marine snails, we don't need to know the details. The complexities sort themselves out, and leave us with a simple tool. To predict the Earth's pull on me, I just need to know how far away the centre of it is, and the mass of the whole planet. The beauty of Newton's theory was that it was simple, it was elegant, and it worked.

But it's still true that forces are weird. In spite of its brilliance, Isaac Newton's explanation of gravity had one major flaw: there was no mechanism. It's straightforward to state that the Earth is pulling on an apple,[1] but what is doing the pulling? Are there invisible strings? Pixies? This wasn't sorted out satisfactorily until Einstein worked out the Theory of General Relativity, but for the 230 years in the middle, Newton's model of gravity was accepted (and is still widely used today) because it worked so incredibly well.

You can't see forces, but almost every kitchen has a device in it for measuring them. That's because you need something important for cooking (and especially for baking) that no glossy recipe book ever mentions. It's necessary because quantities matter: you have to measure 'stuff', and you have to do it accurately. The unmentioned critical ingredient that lets you do this is simple: something (anything) the size of a planet. It's very fortunate for all fans of Eccles cakes, Victoria sponge and chocolate gateau that we're sitting right on top of one.

I've got a book full of hand-written recipes that I've been adding to since I was eight or nine years old, and I love being able to go straight back to the ones of my childhood. Carrot cake is one of those, scribbled on a page smudged with the years, and the recipe starts with an instruction to procure 200 g of plain flour. So the baker does something very clever that we all take completely for granted. They put some flour in a bowl and measure directly how much the Earth is pulling on it. That's what scales do. You put them in the gap between the vast planet and the tiny bowl, and measure the squeeze. The pull between an object and our planet is directly proportional to the mass of both the object and the Earth. Since the mass of the Earth isn't changing, that pull depends solely on the mass of the flour that went into the bowl. Scales measure weight, which is the force between flour and planet. But the weight is just the mass of the flour multiplied by the strength of gravity, which is a constant in our kitchens. So

[1] Yes, I know the story is apocryphal; but the fact is still true!

if you measure the weight and you know the strength of gravity, you can work out the mass of flour in the bowl. Next you need 100 g of butter, so you put butter into the bowl until the squeezing force is half what it was before. This is a fantastically useful and very simple technique for getting at how much stuff you have, and it works for everyone on the planet. Heavy objects are heavy only because they consist of more 'stuff', so Earth is pulling harder on them. Nothing is heavy in space because the local gravity is too weak to pull noticeably on things, unless you're very close to a planet or a star.

But what those kitchen scales are really telling you is that gravity, the grand force that holds together our planet and our solar system and dominates our civilization, is unbelievably weak and weedy. The Earth has a mass of 6×10^{24} kg (6 thousand billion billion tons, if you prefer those units), and it can only pull on your bowl of flour with the force of a small elastic band. It's just as well, otherwise life wouldn't be able to exist, but it does put things into perspective a bit. Every time you pick up an object, you are resisting the gravitational pull of a whole planet. The solar system is large because gravity is weak. Gravity does have one major advantage over all the other fundamental forces, though, and that's its reach. It may be weak, and getting even weaker as you travel further from Earth, but it stretches out across the vast distances of space, tugging on other planets and suns and galaxies. Each tug is tiny, but it's this frail force field that gives our universe its structure.

Still, even picking up the finished carrot cake does take some effort. When it's sitting on the table, the table surface is pushing upwards on the cake just enough to exactly balance out the pull between cake and planet. To pick it up, you have to provide that much force plus a tiny bit more, enough so that the overall force on the cake is upwards. Our lives are controlled not by what individual forces are acting, but on what's left over on balance. And that simplifies things a lot. Massive forces can be made irrelevant by setting them in opposition to other massive forces. The easiest place to start thinking about this is with solid objects, because they keep their shape while they're being pulled. And London's Tower Bridge is very solid indeed.

Gravity can be a nuisance, because sometimes you want to hold things in the air. To do that, you need to resist the downward pull. If you couldn't, everything would slither around on the floor. Fluids flow downwards, and that's just the way it is. For solids, things are different. One single concept, the pivot, lets us effectively neutralize gravity by turning stupidly heavy things into one half of a see-saw. The mysterious other half is often cunningly hidden away, and there's no better example of this than the two graceful towers of Tower Bridge in London. Built on two man-made islands, each a third of the way across the Thames, these towers stand guard at the entrance into London from the sea, and carry the road linking the north of the city to the south.

The pavement is a noisy circus of tourists engaged in camera choreography, while London's taxis, souvenir merchants, coffee stalls, dog-walkers and buses just get on with things in the background. Our tour guide strides through all the chaos and we toddle along behind him like a line of obedient ducklings. He opens an iron gate at the base of one of the towers, ushers us round the corner into a sort of posh garden shed made of stone, and suddenly it's calm. You can almost hear the sigh of relief as his flock realize they've survived the tourist gauntlet and have arrived at their reward: the brass dials, giant levers and reassuringly robust-looking valves of solid Victorian engineering. The pretty and delicate fairy-castle exterior of Tower Bridge is famous around the world, but we are here to see what's lurking inside: the massive steel guts of this elegant and powerful beast.

London has been a port for two thousand years, and the nice thing about having a city on a river is that you have two banks to play with, not just one bit of coastline. But the Thames is both a vital highway for anything that floats and a massive obstacle for anything that walks or rolls. Bridges came and went over the centuries, and by the 1870s the city was crying out for another one. The problem was: how do you satisfy the owner of the horse and cart without cutting off the river to the tall ships? Tower Bridge is the ingenious solution.

The small stone shed squats on top of a spiral staircase that leads downwards into a series of improbably large brick grottos, hidden inside the foundations of the tower. It's like the wardrobe that leads into Narnia, except that this is Narnia for engineers. The first grotto contains the original hydraulic pumps and the next (much bigger) one is mostly filled by a wooden monster: a two-storey-high barrel which used to act as a temporary energy store – a non-electric battery. But it's the third one, the largest of all, that I've really come to see. This is the chamber that houses the counterweight.

The road between the two towers splits into two separate halves. About a thousand times each year, a ship or a boat arrives at the bridge, and the traffic is stopped. Each half of the road swings upwards, and on the other side of the pivot, in this dark chamber underneath the tower, the hidden half of the bridge swings downwards. I look upwards at the underside of this see-saw, and ask what exactly is hanging above us. Glen, our guide, is quite cheery about it. 'Oh, there's about 460 tons of lead ingots and bits of pig iron up there,' he says. 'It rattles around loose – you can hear it when the bridge opens. When they change anything on the bridge, they usually add a bit or take some away so it stays perfectly balanced.' We are apparently standing directly underneath the biggest beanbag in the world.

It's the balance that is the key. Nothing ever lifts the bridge. All those engines do is tilt it a bit – what's on one side of the pivot point is exactly balanced by what's on the other side. This means that very little energy is required to move it – only just enough to overcome the friction of the bearings. Gravity effectively goes away as a problem, because the pull downward on one side is exactly balanced by the pull downward on the other side. We can't beat gravity, but we can use it against itself. And you can make a see-saw as large as you like, as the Victorians recognized.

After the tour, I walked a little distance along the river and then turned round to look back at the bridge. My view of it had completely changed, and I absolutely loved seeing it differently. The Victorians didn't have electricity on tap, computers to control things, or swanky new materials like plastic and reinforced concrete. But they were masters of simple physical principles, and the simplicity of the bridge really gets to me. It's precisely because it's based on something so simple that it's still working after 120 years, with almost no alteration. The gothic revival architecture (which is apparently the technical term for 'fairy-castle style') is just wallpaper covering up a giant see-saw. If they ever build one like this again, I hope that they make some of it transparent, so that everyone can see the genius of it.

This trick for reducing the problems of gravity can be seen all over the place. For example, imagine a pivot point 4 metres above ground with two 6-metre-long halves of a see-saw balancing each other out on either side. This isn't a bridge. This is a *Tyrannosaurus rex*, the iconic carnivore of the Cretaceous world. Two chunky legs hold it up, and the pivot point is at its hips. The reason it didn't spend its life falling flat on its face is that the large heavy head with its terrifying teeth was balanced out by a long, muscular tail. But life as a walking see-saw comes with a problem. Even a very determined *T. rex* would sometimes have needed to change direction, and they were lousy at it. It's been estimated that it could take one or two seconds for them to turn through 45 degrees, making them a bit more cumbersome than the clever, agile *T. rex* of *Jurassic Park*. What could limit a huge, strong dinosaur in such a way? Physics to the rescue . . .

Spinning ice-skaters bring many things to the world – aesthetics, grace and astonishment at what the human body is capable of.

But if you hang around physicists explaining things often enough, you could be forgiven for thinking that an ice-skater's sole contribution is to show everyone that sticking your arms out makes you spin more slowly than when they're tucked in. They're a useful example because ice is more or less free of friction, and so once somebody is spinning, they have a fixed 'amount' of spin. There's nothing to slow them down. So it's really interesting that when they change their shape, they also change their speed. It turns out that when things are further from the axis of spin, they have to travel further on each turn, and so effectively take up more of the available 'spin'.[1] If you stick your arms out, they're further away from the axis, and so the speed of rotation goes down to compensate. And this is basically the problem that the *T. rex* had. It could only generate so much turning force ('torque') with its legs, and because its huge head and tail were sticking out just like very fat, heavy, scaly versions of the skater's arms, it could only turn slowly. Any small agile mammal (for example, one of our very distant ancestors) would be a lot safer once it had worked that out.

The same thought also explains why we put our arms out sideways when we think we're about to fall over. If I'm standing upright and I start to fall to my right, I'm rotating around my ankles. If I stick my arms out or up before I start to fall, the same tipping force won't move me as far, and so I've got more time to make adjustments to stay standing up. That's why gymnasts on the balance beam almost always have their arms out horizontally – it's increasing their moment of inertia, so they've got more time to correct their posture before they fall too far. Having your arms out also lets you rotate yourself by lifting or dropping them, and that helps your balance, too.

In 1876, Maria Spelterina became the only woman ever to cross Niagara Falls on a tightrope. There's a photograph of her

[1] Angular momentum, for the purists.

halfway across, serenely balanced and with peach baskets attached to her feet (to increase the drama). But the most obvious prop in the photo is the long horizontal pole she's carrying, the best aid to balance. Arms will only reach out so far, but this arm-substitute was a large part of the reason for Maria's exquisite control.[1] If she started to lose her balance, it would only happen very slowly, because the distance between the ends of the pole means that the same torque has less effect. Maria was concerned with falling over to one side, but the long pole would also have made it very difficult for her to twist from left to right. And so it was with the *T. rex*. The same bit of physics that was Maria's best defence against falling 50 metres to certain death in the churning water had also, 70 million years earlier, made it impossible for a *T. rex* to change direction quickly.

Gravity pulling on solid objects is a familiar concept, mostly because we're solid objects that are pulled on. But around the solid objects in our world, fluids are flowing – air and water are shifting around in response to the forces acting on them. I think it's a great tragedy that we can't usually see fluids shifting as clearly as we see leaves falling or bridges rising. Liquids experience the same forces, but they aren't limited to holding the same shape, and so the world of fluid dynamics is beautiful: sweeping, whirling, meandering, surprising and everywhere.

The lovely thing about bubbles is that they *are* everywhere. I think of them as the unsung heroes of the physical world, forming and popping in kettles and cakes, bioreactors and baths, doing all sorts of useful things while often only fleetingly in existence. They're such a familiar part of our background that we often don't really see them. A few years ago, I was asking groups of five- to eight-year-old children where you might find bubbles, and they all happily told me about fizzy drinks, baths

---

[1] Subsequently, she crossed with her hands and feet manacled, and also blindfolded.

and aquariums. But the last group of the day were tired, and my cheerful encouragement was met with a grumpy silence and blank stares. After a long pause, and a lot of shuffling, one unimpressed six-year-old stuck his hand up. 'So,' I said brightly, 'where might you find bubbles?' The boy stared at me with a do-I-really-have-to glare, and announced loudly: 'Cheese . . . and snot.' I couldn't fault his logic, although I had never thought about either. It seemed likely that his experience of bubbly snot outranked mine, anyway. But for at least one animal, bubbly snot is the key to a whole lifestyle. Meet the purple marine snail, *janthina janthina*.

The snails that live in the sea generally scoot about on the ocean floor or on rocks. If you were to prise one off its rock, carry it a little way up into the water and let go, it would sink. The ancient Greek polymath Archimedes (he of 'Eureka' fame) was the first to work out the principle that determines when something is going to float and when it's going to sink. He was probably much more interested in ships, but the same principle applies to snails and whales and anything else which is submerged or semi-submerged in any fluid. Archimedes worked out that there is effectively a competition between the submerged object (our snail) and the water that would be there if the snail wasn't. Both the snail and the water around it are being pulled downwards towards the centre of the Earth. Because the water is a fluid, things can move around in it very easily. The gravitational pull on an object is directly proportional to its mass – double the mass of your snail and you double the pull on it. But the water around it is also being pulled downwards, and if the water is pulled down more, the snail will have to float upwards so there's more room for water underneath. Archimedes' Principle, stated for our hapless mollusc, is that there is an upward push on the snail equal to the downward gravitational pull of the water that should have been in that space. This is the buoyancy force, and every submerged object experiences it. In practical terms, it means that if the snail has a greater mass than the water that would fill a snail-shaped hole, it will win the gravitational battle and sink. If the snail has less mass (and is therefore less dense), the water will win the battle to be pulled downwards and the snail will float. Most marine snails are more dense than sea water overall, and so they sink.

For most of their history, marine snails just sank and that's the way it was. But at some point in the past, a 'normal' marine snail had a bad day and got an air bubble trapped in its egg cases. The clever bit about buoyancy is that it's only the average density of the overall object that counts. You don't have to change the mass of the object. You can just change the space it takes up – and air bubbles take up lots of space. One day, a bigger air bubble was trapped, the balance tipped the wrong way, and the first marine snail took flight through the water and drifted up towards the sunlight. The door to the vast larder at the sea surface had been opened . . . but only for a snail that could puff itself up; and so evolution got to work.

Today, *Janthina janthina*, the descendant of the first snails that got lost in space, is common in the warmer oceans of the world. Now bright purple, the snails secrete mucus, the same sort of slime you see on garden stones in the early morning, and use their muscular foot to fold the mucus and trap air from the atmosphere. They build themselves a large bubble raft, often bigger than themselves, to ensure that their total density is always less than the sea water they're in. So they always float, upside down (bubble raft up, shell beneath), preying on passing jellyfish. If you see a purple snail shell on a beach, it's probably from one of these.

Buoyancy can be a very useful and quick indicator of what's inside a sealed object. For example, if you take identically sized cans of a fizzy drink, one diet version and one with a full load of sugar, you'll see that the diet can floats in fresh water and the other one sinks. The cans have exactly the same volume, so the difference is all inside, and it's all that dense sugar. A standard 330 ml can of pop has 35–50 g of sugar inside it, and that extra mass counts, making the can overall more dense than water. That means it beats the water in the battle with gravity, and so it sinks. The mass of sweetener in diet pop is minuscule, so that can is basically just filled with water and air, and it floats. A slightly more useful example is a raw egg. Fresh eggs are more dense than water, so they sink and lie flat in cold water. But if they've been sitting in your fridge for a few days, they'll have been gradually drying out, and as the water sneaks out of the shell, air molecules sneak into an air sac at the rounded end to fill the gap. An egg that's about a week old will sink but stand up on the pointy end (so that the additional air is closer to the surface). And if the egg floats completely, it's been around for a bit too long – have something else for breakfast!

Of course, if you can control the amount of air you're carrying with you, and how much space it takes up, you can choose whether to float or sink. When I first started studying bubbles, I remember finding a paper written in 1962 that stated authoritatively: 'Bubbles are created, not only by breaking waves, but also by decaying matter, fish belchings and methane from the sea-floor.' Fish belchings? It seemed clear to me that this had been written from the blinkered comfort of a large leather armchair, probably in the depths of a London club and much closer to the port decanter than the real world. I thought it was a very funny misconception, and said so. Three years later, while working underwater in Curaçao, I turned around to see a massive tarpon (about 1.5 metres long), swimming just over my shoulder and belching copiously from its gills. That was me told . . . In fact, many bony fish do have an air pocket known as a swim bladder to help them control their buoyancy. If you can keep your density exactly the same as your surroundings, you're in balance and you'll stay put. The tarpon's swim bladders are unusual (tarpon are a rare example of a fish that can breathe air directly as well as extracting oxygen via its gills), but I had to admit that fish do belch. I still maintain that it's not a significant contributor to the number of ocean bubbles, though.[1]

The consequences of gravity depend on what is being pulled on. Tower Bridge is a solid object, and so gravity can change the position of the bridge but not its shape. The snail is also a solid object, and it's moving through ocean water that can flow around it to adjust. But gases can flow too (their ability to flow is why both liquids and gases are called fluids). Solid objects can also move through gases as they follow the pull of gravity: a helium party balloon and a Zeppelin rise for the same reason the bubbly snotty snail does. They are fighting the battle of gravity with the fluid around them and losing.

So the presence of a constant gravitational force can make things unstable, which generally means that there are unbalanced forces and things will shuffle around until balance is restored. If a solid object becomes unstable it flips over or falls down, and any liquid or gas surrounding it will just flow around to make room for the movement. But what happens when the thing that is unstable isn't a single solid object like a balloon, but the fluid itself?

Strike a match, light the wick of a candle, and a fountain of bright, hot gas is switched on. Candle flames have cast a warm glow over scribes, conspirators, schoolchildren and lovers for centuries. Wax is a soft, unassuming fuel, and that makes its transformation all the more surprising. But each one of these

---

[1] When these swim bladders evolved, they provided a huge evolutionary advantage by reducing the energy needed to stay at the same depth. But in recent years they have become a significant disadvantage, because those swim bladders are very easily detectable using acoustics. One of the major technologies that has enabled the vast overfishing of our seas is the 'fish-finder', an acoustic device which is tuned to spot air bubbles and so, by implication, fish. Whole shoals can be chased and wiped out, just because their bubble of air gives them away.

familiar yellow flames is a compact and powerful furnace, fierce enough to smash apart molecules and forge tiny diamonds. And each one is sculpted by gravity.

As you light the wick, the heat of the match melts both the wax in the wick and also the wax close to it, and the first transformation is to liquid. Paraffin waxes are hydrocarbons, long chain molecules with a carbon backbone that's twenty to thirty atoms long. The heat doesn't just give them the energy to slither over each other like a pile of snakes (which is what liquid wax would look like if you could see the molecules). Some will get enough energy to escape completely, drifting out and away from the wick. A column of hot gaseous fuel forms – so hot that it pushes hard on the surrounding air, taking up a huge amount of space for a relatively small number of molecules. The molecules are the same, so the gravitational pull on them is the same in total. But now they're taking up more space, so the gravitational pull for every cubic centimetre has gone down.

Just like a bubbly snotty snail in the ocean, this hot gas must rise because there is cool dense air trying to slide underneath it. The hot air is pushed up an invisible chimney, mixing with oxygen along the way. Even before you've moved the match away from the candle, the fuel is breaking apart and burning in the oxygen, making the gas even hotter. These are the blue parts of the flame, and they reach a staggering 1,400°C. The fountain that you've started intensifies, as the hot air is pushed upwards ever more quickly. It's fed from beneath because the wick is just a long thin sponge, soaking up other wax molecules that have been melted by the furnace.

But the fuel doesn't burn perfectly. If it did, the flame would stay blue and candles would be useless as light sources. As the long chain molecules are snapped and bullied by the heat, some of the detritus remains unburned because there isn't enough oxygen to go round. Soot, tiny specks of carbon, is carried upwards by the flow and heated. This is the source of the comforting yellow light that glows as the soot reaches 1,000°C. The light of a candle is only a byproduct of the fierce heat, and this light is just the glow of a miniature hot coal in a fire. These tiny carbon particles are so hot that spare energy in the form of light is pouring off them, and out into the surroundings. It's been discovered that the maelstrom of a candle doesn't just produce soot in the form of graphite (the stuff we think of as black carbon). It also produces tiny amounts of the more exotic structures that can be formed when carbon atoms join together: buckyballs, carbon nanotubes, and specks of diamond. It's been estimated that the average candle flame produces 1.5 million nanodiamonds each second.

A candle is the perfect example of what happens when a fluid needs to rearrange itself to satisfy the pull of gravity. Hot burning fuel rises very quickly as cool air pushes underneath, forming a continuous convection current. If you blow out the candle, the column of gaseous fuel will keep streaming upwards above the candle for a few seconds, and if you lower a match down from above, you'll see the flame jump to the wick as the column is re-lit.[1]

Convection currents like this help move energy around and share it out wherever a fluid is heated from below. They are why fish-tank heaters, underfloor heating and saucepans on the stove are all so effective – none of those things would work nearly as well without gravity. When we say 'heat rises', that's not quite true. It's more that 'cooler fluid sinks as it wins the gravitational battle'. But no one thanks you for pointing that out.

---

[1] In 1826 Michael Faraday, the famous nineteenth-century experimentalist credited with many practical scientific discoveries, founded a series of talks at the Royal Institution in London, aimed at children, that still continues today – the RI Christmas Lectures. Among his own contributions was a series of six lectures called 'The Chemical History of a Candle', in which he discussed the science of candles, illustrating many important scientific principles that had other applications in the world. I bet he would have been astonished to hear about the nanodiamonds, and probably delighted that the simple candle was still yielding surprises.

Buoyancy doesn't just matter for hot air balloons, snails and romantic candlelit dinners. The oceans, the vast engines of our planet, take their marching orders from gravity just like everything else. The depths are not still. Water that hasn't seen sunlight for centuries is flowing across and around the planet, on a long, slow journey back to the daylight. But before looking down into the depths, look up. Next time you see a tiny moving glint high in the sky on a clear day, a passenger jet at cruising altitude, allow yourself to appreciate just how high up it is: about 10 km. Then imagine yourself standing at the deepest part of the ocean floor, the bottom of the Marianas Trench. The ocean surface would be the same distance above you as that plane is.[1] Even the *average* depth of the oceans is 4 km, a little under half the distance to that plane. The ocean covers 70 per cent of the surface of the Earth. There is a lot of water out there.

And hidden in those dark depths is a familiar pattern. The same mechanism causing raisins to dance in lemonade is also driving the vast oceans of the Earth on their slow journey around the planet. The scale is different and the consequences are more important, but the principle is exactly the same. The blue of our blue planet is in motion.

But why would it move? The oceans have had millions of years to adjust to their situation. Surely they'd have reached wherever they're going by now? Two things keep stirring the pot: heat and salinity. They matter because they affect density, and a fluid with areas of different density will flow to adjust as the battle of gravity plays out. We all know that the ocean is salty, but I am still staggered every time I really think about just how much salt is out there. To make a standard full household bath as salty as the ocean, you need to add about 10 kg of salt, a large bucketful. A

[1] The cruising altitude of a commercial aircraft is about 10,000 metres, and the Challenger Deep, the deepest part of the Marianas Trench, is 10,994 metres deep.

whole bucket, just for one bath! It's not the same everywhere in the ocean – the salinity ranges from about 3.1 per cent to about 3.8 per cent, and although that difference sounds tiny, it matters. Just as putting sugar in a fizzy drink makes it more dense, the huge amount of salt makes sea water more dense than fresh water. Colder water is more dense than warm water, and the oceans range from about 0°C close to the poles to 30°C close to the equator. So cold, salty water will sink and warmer, fresher water will rise. And that simple principle takes sea water on a continual journey around the planet. It may be thousands of years before one bit of water returns to the same part of the ocean again.

In the North Atlantic,[1] water is cooling as the wind steals heat away. Where the sea surface freezes to form sea ice, the new ice is mostly just water; the salt gets left behind. Together, those processes make the sea water colder, saltier and denser, and so it starts to sink, pushing the less dense water out of its way as it answers the call of gravity and finds its way to the bottom of the sea. As it slithers slowly along the sea floor, it's channelled by valleys and blocked by ridges, just like a river. From the North Atlantic, it flows southwards along the bottom of the ocean at a few centimetres per second, and after a thousand years it reaches its first obstacle: Antarctica. Unable to creep further south, it turns to the east as it meets the Southern Ocean. This ocean, the great watery roundabout at the bottom of the planet, links all the sea water on the planet because on its way round the white continent, it merges with the lower edge of the Atlantic, the Indian and the Pacific oceans. The vast, slow flow of water from the North Atlantic creeps around Antarctica until it turns northwards again, journeying far into either the Indian or the Pacific Ocean. Gradual mixing with the water around it reduces its density, and it eventually finds its way back to the surface, after perhaps 1,600 years without a single sunbeam passing through it.

[1] And also close to the coast of Antarctica.

There rain, runoff from rivers and ice-melt dilute the salt again, while wind-driven currents push it along on the rest of its journey until it finds its way back into the North Atlantic, perhaps about to repeat the cycle. It's called the thermohaline circulation: 'thermo' for heat, and 'haline' because of the salt. This ocean overturning is also sometimes referred to as the Ocean Conveyor Belt, and although this conveys a slightly simplistic picture, these flows do girdle the planet and they are driven by gravity. The wind-driven surface currents have carried explorers and traders for centuries. But the ocean conveyor system as a whole carries a cargo of at least equal importance to our civilization: heat.

More heat from the Sun is absorbed at the equator than in any other area on the planet, both because the Sun is higher in the sky there, and because the planet is widest there and so there's a large area for absorption. Heating up water even a tiny bit takes a lot of energy, so warm oceans are like a giant battery for solar energy. The shifting ocean is redistributing that energy around the planet, and the thermohaline circulation is the hidden mechanism behind our weather patterns. Much of our thin, fickle atmosphere whooshes about on top of a steady heat reservoir that constantly provides energy and moderates extremes.

The atmosphere gets all the glory, but the oceans are the power behind the throne. Next time you look at a globe, or a satellite picture of Earth, don't think of the oceans as the empty blue bits between all the interesting continents. Imagine the tug of gravity on those giant, slow currents, and see the blue bits for what they are: the biggest engine on the planet.

3

Small is beautiful

Surface tension and viscosity

COFFEE IS A FANTASTICALLY valuable global commodity, and the precise black magic needed to extract perfection from this humble bean is a constant source of debate (and some snobbishness) for connoisseurs. But my particular interest in it doesn't depend on how it was roasted or the pressure in your espresso machine. I'm fascinated by what happens when you spill it.[1] It's one of those everyday oddities that no one ever questions. A coffee puddle on a hard surface is unremarkable, just a patch of liquid in a blobby shape. But if you leave it to dry, you'll come back to find a brown outline, reminiscent of the line drawn around the body in a detective drama from the 1970s. It was definitely filled in to start with, but during the drying process all the coffee has moved to the outside. Scrutinizing a coffee puddle to see what's going on is the caffeine-waster's equivalent of watching paint dry, but even if you tried, you wouldn't see very much. The physics shunting the coffee around only operates on very small scales, mostly too small for us to see directly. But we can definitely see their consequences.

If you could zoom in on the puddle, you'd see a pool of water molecules playing bumper cars, and much bigger spherical brown particles of coffee drifting around in the middle of the

---

[1] Sorry. Really. If it helps, what I'm about to say is just as true for instant coffee, so you don't ever need to waste shots of fancy coffee on science.

game. The water molecules attract each other very strongly and so if a single molecule lifts up a bit from the surface, it immediately gets pulled back down to join the horde below. This means that the water surface behaves something like an elastic sheet, pulling inward on the water below it so that the surface is always smooth. This apparent elasticity of the surface is known as surface tension (of which much more a little later). At the edges of the puddle, the water surface curves downward smoothly to meet the table, holding the puddle in place. But the room is probably warm, and every so often a water molecule escapes from the surface completely and floats off into the air as water vapour. This is evaporation, and it happens gradually, and only to the water molecules. The coffee can't evaporate, so it's effectively trapped in the puddle.

The clever bit happens as more and more water escapes, because the water edge is pinned to the table (we'll see why later). Water is so strongly stuck to the table that the edge has to stay where it is. But evaporation is happening at the edges more quickly than from the middle, because a higher proportion of the water molecules are exposed to the air there. The bit that you can't see (as you try to persuade your coffee companion that watching paint dry really is the latest thing) is that the contents of the puddle are on the move. Liquid coffee from the middle must flow out to the edges to replace the lost water. The water molecules carry the coffee particles along as passengers, but when it's their turn to escape into the air, the coffee can't join them. So the coffee particles are gradually carried out to the edges, and once the water has completely gone, all that is left is a ring of abandoned coffee.

The reason I find this so fascinating is that it happens right in front of your nose, but all the interesting bits are just too small to watch properly. This world of the small is almost a whole different place. The rules that matter are different down there. As we'll see, the forces we're used to, like gravity, are still present.

But other forces, the ones that arise because of the way molecules dance around each other, start to matter more. When you dive down into the world of the small, things can seem very weird. It turns out that the rules that operate on this small scale explain all sorts of things in our larger-scale world: why there's no cream on the milk any more, why mirrors fog up, and how trees drink. But we're also learning to use those rules to engineer our world, and we'll see how they're going to help us save millions of lives through improved hospital design and new medical tests.

*

Before you can worry about things that are too small to see, you have to know that they're there. Humanity faced a catch-22 here – if you don't know there's anything there, why would you go looking for it? But all of that changed in 1665 with the publication of one book, the first scientific bestseller: Robert Hooke's *Micrographia*.

Robert Hooke was the Curator of Experiments at the Royal Society, and so he was a generalist, free to roam among the scientific toys of the day. *Micrographia* was a showcase for the microscope, designed to impress the reader with the potential of this novel device. The timing was perfect. This was an era of great experimentation and rapid advances in scientific understanding. Lenses had been pootling around the edges of human civilization for a few centuries, mostly unappreciated and seen as novelties rather than serious tools for science. But with *Micrographia*, their moment had arrived.

The wonderful thing about this book is that although it wears the robes of respectability and authority, as befits a publication of the Royal Society, it's unashamedly the product of a scientist at play. It's full of detailed descriptions and beautiful illustrations, expensively produced and carefully presented. But underneath all that, Robert Hooke was basically doing what every child does when given a microscope for the first time. He just went round looking at everything. There are stunningly detailed pictures of razorblades and nettle stings, grains of sand and burnt vegetables, hair and sparks and fish and bookworms and silk. The level of detail revealed in this tiny world was shocking. Who knew that a fly's eye was so beautiful? In spite of the careful observations, Hooke didn't make any claims to in-depth study. In the section on 'gravel in urine' (the crystals commonly observed on the insides of urinals), he speculates on a way of curing this painful affliction and then happily leaves the hard work of actually solving the problem to someone else:

> It may therefore, perhaps, be worthy some Physicians enquiry, whether there may not be something mixt with the Urine in which the Gravel or Stone lies, which may again make it dissolve it, the first of which seems by it's regular Figures to have been sometimes *Crystalliz'd* out of it . . . But leaving these inquiries to Physicians or Chymists, to whom it does more properly belong, I shall proceed.

And proceed he does, dancing through mould and feathers and seaweed, the teeth of a snail and the sting of a bee. On the way, he coins the word 'cell' as a description of the units that made up cork bark, marking the start of biology as a distinct discipline.

Hooke hadn't just shown the way to the world of the small; he'd thrown open the doors and invited everyone in for a party. *Micrographia* inspired some of the most famous microscopists of the following centuries, and also whetted the scientific appetite of fashionable London. And the fascination came from the fact that this fabulous bounty had been there all along. The annoying black speck buzzing around rotting meat was now revealed as a minute monster with hairy legs, bulbous eyes, bristles and shiny armour. It was a shocking discovery. By then, great voyages had crossed the world, new lands and new people had been discovered, and there was great excitement about what was to be found in far-away places. It hadn't really occurred to anyone that navel-gazing might have been severely underrated, and that even belly-button fluff might have much to say about the world. And once you'd got over the shock of the flea's hairy legs, you could see how they worked. The world down there was mechanical, it was comprehensible, and the microscope made sense of things that humans had noticed for years but hadn't been able to explain.

But even that was just the start of the voyage into the world of the small. Over two centuries more would pass before the existence of atoms was confirmed, each one so tiny that you'd need 100,000 of them to make a line as long as a single one of the cork cells. As the famous physicist Richard Feynman was to point out many years later, there's plenty of room at the bottom. We humans are just lumbering about in the middle of the size scales, oblivious to the minuscule structures that our world is built of and built on. But 350 years after the publication of Hooke's *Micrographia*, things are changing. We can do more than just peer into that world like a child peering into a protective glass museum case, forbidden to touch. Now we're learning to manipulate atoms and molecules on that scale; the glass is off the case, and we can join in. 'Nano' is in fashion.

A major part of what makes the world of the tiny both fascinating and extremely useful is that things work differently at that level. Something that's impossible for a human might be an essential life skill for a flea. All the same laws of physics apply; the flea exists in the same physical universe as you and me. But different forces take priority.[1] Up here in our world, there are

---

[1] We can go a long way into the world of the small without having to deal with the strangeness of quantum mechanics. That really kicks in when you explore what's happening to individual atoms and molecules, and there's an awful lot that's bigger than that and still smaller than what we can see. That middle bit is interesting because we can understand it intuitively (something

two dominant influences. The first is gravity, pulling us all downwards. The second is inertia; because we're so big, it takes a lot of force to get us moving or to slow us down. But as you get smaller, gravitational pull and inertia also get smaller. And then they find themselves in competition with the other weaker forces that were there all along, but insignificant. There's surface tension, the force shifting the coffee granules about as the coffee puddle dries. And then there's viscosity. Viscosity in the world of the small is why we don't get a nice layer of cream on top of the milk any more.

It was always the gold- and silver-topped milk bottles they went for. If you were early enough, and careful when you opened the front door, you'd catch them at it. Bright-eyed perky little birds, perched on the top of the bottle, snatching hasty sips of cream from the hole they'd pecked in the thin aluminium bottle top while keeping a beady eye on the world around them. As soon as they knew they were rumbled, they were off, probably to try their luck at the neighbour's doorstep. For about fifty years in this country, blue tits were masters at stealing cream. They learned from each other that right below the flimsy lid there was a rich fatty treasure, and that knowledge spread throughout the UK blue-tit population. Other bird species didn't seem to cotton on to this trick, but the blue tits were waiting for the milkman every morning. And then the game came to an end quite suddenly, not just because of plastic milk bottles but because of something more fundamental. For as long as humans have been milking cows, the cream has risen to the top. But these days, it doesn't.

The bottle that the hungry blue tit was hopping about on contained a mixture of all sorts of goodies. Most of milk (nearly 90 per cent) is water, but floating around in that are sugars (that's

---

which is impossible by definition when it comes to the rules of the quantum world), even though we can't see it clearly.

the lactose that some people can't tolerate), protein molecules assembled into minuscule round cages, and bigger globules of fat. All of this is jumbled up together, but if you leave it to sit for a while, a pattern emerges. The fat globules in milk are tiny – between 1 and 10 microns in size, which means you could fit somewhere between 100 and 1,000 of them in a line between the millimetre markers on a ruler. And those tiny blobs are less dense than the water around them. There's less 'stuff' in the same volume of space. So as they're being jostled about with everything else, there's a tiny difference in where they go. Gravity is pulling the water around them downwards a tiny bit harder than it's pulling the fat globules, and the fat is very gently squeezed upwards. That means that the fat is ever so slightly buoyant, and will very slowly rise up through the milk.

The question is: how fast will it rise? And here's where the viscosity of the water starts to matter. Viscosity is just a measure of how hard it is for one layer of a fluid to slide over another layer. Imagine stirring a cup of tea. As the spoon goes round, the liquid around the spoon has to move, flowing past other liquid next to it. Water isn't very viscous, so it's very easy for those layers to move past each other. But then think about stirring a cup of syrup. Each sugar molecule is holding on to the ones around it very firmly. To move these molecules past each other, you've got to break those bonds before the molecules can move on. So it's hard work to shunt the fluid about, and we say that the syrup is viscous.

In the milk, the fat globules are pushed upwards because they're buoyant. But if they want actually to move upwards, they have to shove the liquid around them out of the way. As part of that pushing process, the nearby liquid has to slide over itself, so its viscosity matters. The more viscous it is, the more resistance there is to the fat globules rising.

Right under the blue tit's feet, this battle is going on. Each fat globule is being pushed upwards by its buoyancy, but it experiences a drag force because of the liquid around it having to move to let it pass. And the same forces acting on the same sort of fat globule come to a different compromise for different globule sizes. The drag has a much greater effect when you're small, because you have a large surface area relative to your mass. You've only got a small buoyancy to use to shove quite a lot of the surrounding stuff out of the way. So even though the smaller fat globule is in exactly the same liquid, it rises more slowly than a bigger one. In the world of the small, viscosity generally trumps gravity. Things move slowly. And your exact size matters a lot.

In the milk, the larger fat globules rise faster, bump into some smaller, slower ones and stick to them, forming clusters. These clusters experience less drag for their buoyancy because they're even bigger than individual globules, so they rise even faster. The blue tit just has to sit and wait at the top, and breakfast will arrive at its feet.

And then came homogenization.[1] Milk manufacturers worked out that if they squeezed the milk at very high pressure through very tiny tubes, they could break up the fat globules and reduce their diameter by about a factor of five. That reduces the mass of each one by a factor of 125. Now the weedy upward buoyant push on each globule provided by gravity is completely overwhelmed by viscous forces. The homogenized fat globules rise so slowly that they might as well not bother.[2] Just making them smaller shifts the battle into different terrain where viscosity can

---

[1] As someone who loves the variety and spice of life, I'm always a bit sad when I see this word. Making everything the same definitely has its uses, but sometimes it does just sound as though it's taking the fun out of life. Especially if you're a blue tit.

[2] Their rise is slowed even more by the extra protein coat that surrounds each of the new smaller globules; this weighs them down a bit, so they're even less buoyant than they were before. This has been measured in quite a lot of detail. You'd be surprised at how much science has gone into a pint of milk.

score a clear victory. Cream won't rise to the top any more. The blue tits had to find another source of breakfast.

So the forces are the same, but the hierarchy is different.[1] Both gases and liquids have viscosity – even though gas molecules don't stick to each other like the ones in liquids, they jostle each other a lot, and the giant game of bumper cars has the same effect. This is why an insect and a cannon ball don't fall at the same speed unless you take all the air away and drop them in a vacuum. Air viscosity matters a lot for the insect, and hardly at all for the cannon ball. If you take the air away, gravity is the only force that matters in both cases. And a tiny insect trying to fly in air uses the same techniques that we use to swim in water. Viscosity dominates their surroundings, just as it does ours in the pool. The smallest insects are swimming through air much more than they're flying through it.

Homogenized milk demonstrates the principle, but its application goes far beyond the doorstep. Next time you sneeze, you might want to think about the size of the droplets you're spraying round the room. What stops cream going up also stops disease coming down.

Tuberculosis has been with humans for millennia. The earliest record of it is in ancient Egyptian mummies from 2400 BC; Hippocrates knew it as 'phthisis' in 240 BC, and European royalty were called upon to cure the 'king's evil' in medieval times.

---

[1] If you're interested in reading more on this, the biologist J. B. S. Haldane wrote a very famous short essay in the 1920s called 'On being the right size'. It's here: http://irl.cs.ucla.edu/papers/right-size.html. The most memorable quote from it is very painfully true: 'To the mouse and any smaller animal it [gravity] presents practically no dangers. You can drop a mouse down a thousand-yard mine shaft; and, on arriving at the bottom, it gets a slight shock and walks away, provided that the ground is fairly soft. A rat is killed, a man is broken, a horse splashes.' As far as I'm aware, no one has actually done this specific experiment. Please don't be the first. And certainly don't blame me if you do.

As the Industrial Revolution drove people to live in towns, 'consumption', the disease of the urban poor, was responsible for a quarter of all deaths in England and Wales in the 1840s. But it wasn't until 1882 that the culprit was found, a tiny bacterium called *Mycobacterium tuberculosis*. Charles Dickens described the common sight of consumptives coughing, but he couldn't write about one of the most important aspects of the malady, because he couldn't see it. Tuberculosis is an airborne disease. Carried out of the lungs with each cough are thousands of fluid droplets, plumes of minuscule crusaders. Some of them will contain the tiny rod-shaped TB bacteria, each only three-thousandths of a millimetre long. The fluid droplets themselves start off fairly big, perhaps a few tenths of a millimetre. These droplets are being pulled downwards by gravity and once they hit the floor, at least they're not going anywhere else. But it doesn't happen quickly, because it's not just liquids that are viscous. Air is too – it has to be pushed out of the way as things move through it. As the droplets drift downwards, they are bumped and jostled by air molecules that slow their descent. Just as the cream rises slowly through viscous milk to the top of the bottle, these droplets are on course to slide through the viscous air to reach the floor.

Except they don't. Most of that droplet is water, and in the first few seconds in the outside air, that water evaporates. What was a droplet big enough for gravity to pull it through the viscous air now becomes a mere speck, a shadow of its former self. If it was originally a droplet of spit with a tuberculosis bacterium floating about in it, it's now a tuberculosis bacterium neatly packaged up in some leftover organic crud. The gravitational pull on this new parcel is no match for the buffeting of the air. Wherever the air goes, the bacterium goes. Like the miniaturized fat droplets in today's homogenized milk, it's just a passenger. And if it lands in a person with a weak immune system, it might start a new colony, growing slowly until new bacteria are ready to be coughed out all over again.

Tuberculosis is treatable if the right drugs are available. That's why it has mostly disappeared from the western world. But at the time of writing, TB is still the second greatest killer of our species after HIV/AIDS, and it's a gigantic problem in the developing world. Nine million people developed TB in 2013, and 1.5 million of them died. The bacterium has changed in response to antibiotics, becoming resistant to so many waves of drugs that it's obvious it can't be eradicated using medicine alone. The number of multi-drug-resistant strains of TB is on the increase. Outbreaks are popping up in hospitals and schools. So recently the focus has shifted to those tiny droplets. Rather than cure TB once you have it, how about changing your buildings to prevent the spread of those disease-laden plumes so it never gets to you in the first place?

Professor Cath Noakes works in civil engineering at the University of Leeds, and she is one of the researchers chipping away at this particular coalface. Cath is very enthusiastic about the potential for relatively simple solutions to emerge from a sophisticated understanding of tiny floating particles. Engineers like her are now learning how these tiny vehicles for disease travel, and it turns out this has very little to do with what's in them and how long they've been there. It has everything to do with the battle of forces on the particle, and the battle lines are drawn by the particle size. It's been discovered that even the larger droplets can travel further than anyone had thought, because turbulence in the air can keep them aloft.[1] The tiniest ones can stay in the air for days, although ultraviolet and blue light damage them. If you know where your particle sits on the size scale, you can work out where it's going to go. So, if you are designing

---

[1] If you keep stirring your milk, the cream won't rise to the top because it keeps getting mixed back in. The same principle applies here – particles don't sink down very far because they keep getting mixed back in by air currents that move faster than they're falling.

a ventilation system for a hospital, it's becoming possible to plan to remove or contain specific particle sizes, and therefore control the spread of disease. Cath tells me that each airborne disease may require a different plan of attack, depending on how much of it you need to get sick (in the case of measles, very little) and where in your body the disease settles (the TB bacterium has different effects in your lungs and your windpipe). These studies are still in their early days, but they're advancing very quickly.

Humans have been at the mercy of TB for generations, but now we can visualize its spread, and that gives us the chance to control it. Where our ancestors saw only a foul room of sickness, awash with mysterious miasmas, we now understand the subtle swirling of the air around each patient, the sorting and shunting of disease particles, and how the consequences take effect. The outcomes of this research will be incorporated into the hospital designs of the future, and many lives will be saved by engineering on the macro scale to influence particles on the micro scale.

Viscosity matters when something small is moving through a single fluid – fat globules rising through milk or a tiny virus falling through the air. Surface tension, its partner in the world of the small, matters at the place where two different fluids touch. For us, that's usually where air touches water, and everyone's favourite example of air mixing with water is a bubble.[1] So let's start with a bubble bath.

The sound of a bathtub filling up is distinctive and jolly. It announces the imminent reward after a hard day, a soak to recover from a particularly tough badminton match or just a bit of pampering. But the moment you pour in a bit of bubble bath, the sound changes. The deep rumble gets softer and quieter as the foam builds up, and the place where the water stops and the air starts gets harder to identify. Pockets of air are trapped inside

---

[1] Especially mine. I am a bubble physicist, after all.

watery cages, and all it took was a tiny amount of stuff from a bottle.

It was a group of European scientists in the late nineteenth century who picked apart the puzzle of surface tension. The Victorians loved bubbles. Soap production expanded dramatically between 1800 and 1900, and the white suds washed the workers of the Industrial Revolution. Bubbles provided the Victorians with good fodder for moralizing; they were the perfect symbol of pure cleanliness and innocence. And they were also a nice example of classical physics at work, just a few years before Special Relativity and quantum mechanics came along and poked a sharp pin in the ballooning idea of a neat, tidy and well-behaved universe. But even so, the serious men with top hats and beards didn't work out the secrets of bubble science all by themselves. Bubbles were so universal that anyone could have a go. Enter Agnes Pockels, often described as a mere 'German housewife', but really a sharp-minded critical thinker, who used the limited materials available to her and a decent dose of ingenuity to examine surface tension for herself.

Born in 1862 in Venice, Agnes was of a generation very firmly convinced that the woman's place was in the home. So that's where she stayed while her brother went off to university to study. But she learned advanced physics from the material he sent to her, carried out her own experiments at home, and generally kept up with what was going on in the academic world. When she heard that the famous British physicist Lord Rayleigh was starting to take an interest in surface tension, something she had done many experiments on, she wrote to him. He was so impressed with the letter describing her results that he sent it to be published in the journal *Nature* so that it could be seen by all the greatest scientific thinkers of the day.

Agnes had done something very simple and very clever. She had suspended a small metal disc (something about the size of a button) on the end of a string and let it sit on the surface of the water. Then she had measured how much force it took to pull it away from the surface. The mystery was that the water held on to the disc; you had to pull harder to get it off the water surface than you would have had to pull to pick it up off the table. That pull from the water is what we call surface tension; so in measuring the pull Agnes was measuring surface tension. She could then investigate the surface of the water, even though the thin layer of molecules responsible for the pull was far too small for her to see directly. We'll see how in a minute; but first, back to the bath.

A bath full of pure water is a jostling swarm of water molecules playing a very crowded game of bumper cars. But one of the things that makes water such a special liquid is that all those molecules are very strongly attracted to all the other water molecules around them. Each one has a larger oxygen atom and two small hydrogen atoms (that's the two Hs and an O in H_2O). The oxygen sits in the middle with the two hydrogens stuck to it on either side, making a shallow V-shape. But although the oxygen is very strongly attracted to and bonded to its own two hydrogen atoms, it's also flirting with any others that happen to go by. So it's constantly tugging on the hydrogen from other water molecules. This is what holds water together. It's called hydrogen bonding, and it's very strong. In the bath, water molecules are constantly pulling on the other water molecules around them, tugging the whole mass of water together.

The water molecules on the surface are a bit left out. They are being pulled by the water molecules underneath them, but there's nothing above them to pull back. So they're being pulled down and sideways but not up: and the effect of this is to make the surface behave like an elastic sheet, pulled tight over all the water molecules below the top layer, and pulling itself inward so that it is as small as possible. This is surface tension.

As you run the tap, air gets carried downward into the bath, making bubbles. But when these bubbles float up to the surface, they can't last. The round dome of the bubble is stretching the surface and the surface tension isn't strong enough to haul it back. So the bubbles burst.

One of the things that Agnes did was to set up her button so that it was being pulled upwards, but not quite hard enough to pull it off the surface. And then she touched the surface of the water nearby with a drop of something like detergent. After a second or so, the button would pop off the surface. The detergent had spread across the water, and it had reduced the surface tension. All it takes to reduce the surface tension is to provide a thin top layer so that the water molecules don't have to be the ones right at the surface.

When it's finally time to add the bubble bath, it's time to say goodbye to a clean, flat, minimal surface. That dollop of scented gloop gets carried down into the water and immediately does its best to hide at the edges. Each molecule has one end that loves water and one end that hates water. If the end that hates water can find some air, it will stay with it, but the water-loving end isn't giving in either. So at any place where water touches air, a thin layer of bubble bath sits right at that surface. It's just one molecule thick, and each molecule is the same way round so that the water-loving ends are all still submerged in water, and the water-hating ends are all still in air. With this thin coating, a large surface isn't a problem. The bubble bath doesn't provide the strong pull that water does, so the elastic sheet effect becomes really weak. It's time for a surface party, and that's what the foam is. By reducing the surface tension, bubble bath makes it easier for bubbles to last because their large surface is much more stable.

It's probably worth noting here that we associate the white foam with things getting cleaner, but in modern detergents the best stuff for sticking to the water surface and making the foam is not the best stuff for pulling dirt and grease off clothes and plates. You can make a very good cleaning detergent which hardly makes foam at all, and in fact the foam often gets in the way. But the purveyors of cleaning products did such a good job of convincing people that beautiful white foam was your guarantee of a thorough cleaning that they've now backed themselves into a corner. Foaming agents are now added to make sure there are bubbles, because otherwise consumers complain.

Like viscosity, surface tension is something that we're aware of up here at our size scales, although it's usually less important than gravity and inertia. As you get smaller, surface tension pushes its way up the hierarchy of forces. It explains why goggles fog up and how towels work. And the real beauty of the world of the small is that you can contain many tiny processes inside one giant object, and their effects add up. For example, it turns out that surface tension, which only dominates in the tiniest of situations, also makes possible the largest living things on our planet. But to get there, we need to look at another aspect of surface tension. What happens when the surface separating a gas and a liquid bumps up against a solid?

My first open-water swim turned out not to be for the faint-hearted. Fortunately, I didn't know that beforehand so I couldn't worry about it. When I was working at the Scripps Institution of Oceanography in San Diego, the big annual event for my swim team was from La Jolla beach to Scripps pier and back, 4.5 km across a fairly deep marine canyon. I had only ever swum properly in swimming pools, but I'm always up for trying something new and I'd been swimming a lot, so I turned up and hoped I didn't look like too much of a rookie. The mass entry to the water was a bit of a scrum, but it got better after that. The first part of the swim was across the top of a stunning kelp forest, and it was almost like flying. The sun glinted through the huge stalks of bull kelp just like it does in forests on land, and then the kelp disappeared downward into the murky depths, reminding me that there were many creatures swimming about down there that I couldn't see. Once we were past the kelp, the water got choppy, and I had to pay much more attention to where we were going. And that was getting harder. The pier was fuzzy on the horizon, and I couldn't see anything at all down below. After slightly too long, I realized that the reason everything had disappeared was that my goggles had fogged up. Oh.

Inside my plastic goggles, sweat had evaporated from the warm skin around my eyes. The harder I worked, the more evaporated. The air trapped between me and my goggles was now a mini-sauna, hot and humid. But the ocean around me was nice and cool, and so my goggles were being cooled from the outside. When water molecules in the air bumped into the nice cool plastic, they gave up their heat and condensed, becoming liquid again. But that wasn't the problem. The problem was that as all those water molecules found each other on the inside of my goggles, they stuck together, far more attracted to each other than they were to the plastic. Surface tension was pulling them inwards, forcing them to collect in tiny droplets so that there was as little surface as possible. Each droplet was tiny – perhaps 10–50 microns across. So gravity was insignificant compared with the surface forces sticking them to the plastic and there was no point in waiting around in the hope of them falling off.

Each little droplet acted like a lens, bending and reflecting the light that hit it. When I lifted my head to look for the pier, the light that had been travelling straight to my eyes was messed up by the droplets. Like a tiny house of mirrors, they had scrambled the image so that I was just looking at vague grey fuzz. I stopped briefly to rinse out my goggles, and for a while I had a crystal clear view of the pier again. But the fog came back. Rinse. Fog. Rinse. Eventually I just stuck next to my swimming partner because she had a bright red swimming cap, and the red made it through the silly little water droplets.

When we reached the pier, we paused to check that everyone was OK. With a bit of time to think, I finally remembered something I'd been taught just a week or so before by a scuba diver. Spit in your goggles, and rub it over the inside of the plastic. At the time, I'd made a face, but now I didn't want to go all the way back across the canyon blind, so I spat. And the swim back was a completely different experience. That was partly because my swim partner had decided that she was bored and wanted it all done with, and I had to struggle to keep up. But it was mostly because I could see – swimmers, kelp, the beach we were aiming for, the occasional curious fish. Human saliva acts a bit like detergent: it reduces surface tension. My goggles were still a mini-sauna and the water was still condensing, but surface tension wasn't strong enough to bunch it up in droplets. So it was just spread out in a thin film covering the entire surface. Since there were no watery lumps and bumps and boundaries, light could travel through in a straight line, and I could see clearly. Back at the beach, I stumbled out of the water euphoric, partly with relief for having finished the swim, and partly with a new appreciation of what the underwater world had to offer.

This is one way to stop things fogging up: to spread a thin layer of surfactant on the surface. Lots of things will do that job – saliva, shampoo, shaving cream or expensive commercial antifog. If the surfactant is ready and waiting, any water that condenses will immediately be coated in it. By providing that coating, you are weakening the surface tension, and swaying the battle of forces in each fog droplet so that the water covers the plastic evenly. The water can stick to the whole surface of the goggles, as long as there are no stronger forces to pull it away. Surface tension is the only other force that stands a chance of competing, so when you weaken that, the problem vanishes.[1]

So one solution is to reduce the surface tension. But there is

[1] You can see this effect for yourself if you put a droplet of water on something that is fairly hydrophobic – a tomato does the job. The droplet will sit up, mostly off the surface. Then just touch it with a cocktail stick with a tiny bit of

another solution: increasing the attractiveness of the goggles. A droplet on its own will suck itself up into a ball. If you put it on plastic or glass, it will sit up high and barely touch since the water molecules will shuffle about until as few as possible touch the plastic. But if you put the droplet on a solid surface that attracts water molecules nearly as strongly as other water molecules do, the water will snuggle up to that surface. Instead of a perky near-spherical droplet, you get a flattened drip that feels the pull of the surface as much as it feels the pull of its neighbours. These days, I buy goggles that have a coating on the inside that attracts water – it's called hydrophilic. Water still condenses, but it spreads out along that surface, attracted to the coating. Condensation in goggles is here to stay, but fogging up is a thing of the past.[1]

Weakening surface tension has its uses. But that pull between individual water molecules is really strong. And the smaller the volume of water you're interested in, the more it matters. So what surface tension is really useful for is plumbing on the tiniest scales. Down there, you don't need pumps and siphons and huge amounts of energy to shunt water around; you just need to make things small enough for gravity to be irrelevant and let surface tension get on with the hard work. Mopping up is boring, but the world would be very different without it.

_____

detergent on the end, and the drop will immediately spread out sideways. I recommend washing the detergent off the tomato before eating it.

[1] This balance – how much water is attracted to a solid surface compared with how much it's attracted to itself – helps with all sorts of problems. The most important one for any Brit is the question of why some teapots dribble from the spout as you finish pouring, sending tea down the side of the pot and on to the table rather than into the cup. The answer is that the teapot is just too attractive to water. As the flow slows, the forces sticking the water to the spout dominate over the momentum carrying the water forwards. You can solve this by having a hydrophobic teapot, one that doesn't attract the tea at all. Sadly, at the time of writing, no one seems to be selling them.

I'm a messy cook, reasonably competent, but far more interested in the cooking process itself than the trail of devastation that I tend to leave behind me. This makes me nervous when using other people's kitchens. Years ago in Poland, I set out to make apple pie for the international group of volunteers I was working with at a school.[1] It didn't start well. The tall, fierce school cook bellowed 'NO!' with some enthusiasm when I asked whether I could use the kitchen, and it took me a few puzzled seconds to remember that we'd been talking in Polish and 'no' is their word for 'yeah'. My Polish wasn't very good, and I didn't follow all of the details that came next, but I took away the very strong message that the kitchen was to be left clean. Very clean. No spilling anything. Definitely immaculate. So later that evening, after she'd gone home and I'd assembled all the ingredients, of course the first thing I did was to knock over a large and newly opened carton of milk.

My first reaction was just to will the milk to vanish, so that the stern cook would never know it had existed. Milk is slippery and sticky, can't be picked up or swept away, and this particular batch of it was advancing along the kitchen floor at an alarming rate. But there is a tool for gathering a liquid together, for putting it all back in one place. It's called a towel.

As soon as the towel touched the milk, the liquid had a new set of forces bossing it about. Towels are made of cotton, and cotton attracts water. Down there on a tiny scale, water molecules were attaching themselves to cotton fibres, and slowly creeping over the surfaces of each fibre. And water molecules are so

---

[1] It was actually an apology. On a trip to Krakow, I had promised a fabulous dinner in the Jewish quarter of the city, but this was before the days of smartphones, so I got lost. I led twelve hungry people on a merry dance around many dark and empty streets, failing to find any restaurants at all, never mind the excellent ones I was aiming for. We ended up eating in a McDonalds instead. I felt that apple pie was the least I could do to make up for this.

strongly attracted to each other that the first one to touch the towel can't crawl upwards by itself. It can only move up if it brings the next water molecule with it. And that one has to bring the following one. So water creeps up the cotton fibres, bringing everything else in the milk with it. The forces sticking the water to the towel fibres are so strong that the measly downward pull of gravity becomes totally irrelevant. What went down happily goes back up.

But this is only half the story. The real genius of a towel is its fluffiness. If a towel could only coat each of its fibres in a thin layer of water, it wouldn't be able to collect together much liquid at all. But the fluff gives the towel lots of air pockets and narrow channels. Once water finds its way into a narrow channel, it's being pulled upwards on all sides, and the water in the middle just gets dragged along as well. The narrower the channel is, the more surface there is for each drip of water in the middle. Fluffy towels have loads of surface, and very narrow gaps in between, so they can suck up a lot of water.

As I watched the puddle of milk disappear into the towel, tiny water molecules were crowded together, jostling inside the fluff. The ones at the bottom were just going along with the crowd, sticking to the other water molecules next to them. The ones touching the cotton were clinging on to both the cotton and the water molecules on the other side, holding their position. The ones touching dry towel were latching on to the new dry cotton and, once they were attached, pulling others up behind them, filling the gaps in the structure. The ones at the surface were tugging on the water molecules directly below them, trying to surround themselves with as many other water molecules as possible, and pulling water upwards in the process. This is capillary action. Gravity was pulling down on all of that milk, wherever it was in the fluff. But gravity couldn't compete with the forces holding the entire thing up, the ones at the top, where milk just touched dry cotton inside millions of tiny air pockets.

As I turned the towel over and moved it around, different regions of the towel filled up, storing water in the pockets.

Water will keep creeping upwards through the gaps, bringing other water with it, until the sum of those tiny forces from a multitude of pockets is finally balanced by the pull of the planet. This is why when you dip the edge of a towel in water, the liquid will spread quickly upward for a few centimetres and then stop. At that point, the weight of the water is exactly balanced by the upward pull of the surface tension. The narrower the channels in the fluff, the more surface there is to contribute surface tension, and so the higher the waterline will be. Scale really matters here – if you made fluff that had the same shape but was a hundred times bigger, it wouldn't be absorbent at all. But when you shrink the shape, you shift the hierarchy of forces and up the water goes.

The best bit of the whole thing is that if you leave the towel out to dry, water will evaporate from those pockets and disappear into thin air. As a means of getting rid of a problem, that's hard to beat; the towel collects and holds on to the liquid until it floats off of its own accord.[1]

The spill vanquished, I finished the apple pie and left the kitchen in a suitably immaculate state. But I had one final problem stored up that no amount of surface science could have helped me with. The whipped cream that I served with the apple pie was thoroughly unpleasant, as the faces of the pie consumers made pretty clear. It wasn't the best way to learn the Polish word for 'sour', the one that preceded the word 'cream' on the pot. Still, you live and learn, and I won't make that mistake again.

The reason why towels are made of cotton is that cotton is mostly cellulose, long chains of sugars that water molecules stick to very easily. Cotton wool, kitchen towel, cheap paper: all of these are absorbent because they have a fluffy structure on a

[1] Of course, the fat and protein and sugar in the milk don't evaporate, so they're left behind and the towel still needs a wash.

tiny scale, made out of water-loving cellulose. The question is: what are the limits of this size-dependent physics? If you make the channels as small as physically possible, what can you do with them? It's not just towels that suck water up tiny channels made of cellulose. Nature got there long before us. The mightiest example of what the physics of the small is capable of is also the largest living organism on our planet: the giant redwood.

*

The forest is quiet and humid. It feels as though it must always have been like this, as though change is rare here. The forest floor between the tree trunks is carpeted with moss and ferns, and the only sounds are the songs of unseen birds and the deep, unnerving creaking as the trees shift their weight. Up above, blue skies are visible between the spindly green branches, and down at my feet there is water everywhere: streams, patches of damp soil, exploratory rivulets on their way down the valley. Every so often as I walk, my subconscious kicks me into alertness, because there's a looming patch of darkness in the forest, something that doesn't fit. But it isn't a predator. It's a tree: one of the real giants, a thousand-year-old colossus lurking among the youngsters, stamping its status on the forest with its shadow.

The coastal redwood, *Sequoia sempervirens*, used to cover vast swathes of this part of northern California. These days, those huge forests have been reduced to a few small pockets, and I'm visiting one of the most well known, the Redwood National Park in Humboldt County. These giants are striking because each tree trunk is completely straight and vertical, reaching only for the sky. The tallest known tree on the planet is here, and it's a staggering 116 metres high.[1] On my hike, I frequently pass

[1] The clock tower of Westminster, the one that houses Big Ben, is 96 metres tall. These trees really are giants.

trees with trunk diameters of 2 metres or more. Possibly most astonishing of all is that just behind the deep ridges and wrinkles in the bark, these trees are still growing new rings. They're alive. The tiny evergreen leaves 100 metres above me are capturing the Sun's energy, storing it up, making the stuff from which new tree is built.

But life demands water, and the water is down here, where I am. So all around me in the forest, water is flowing upwards. And this flow has never been interrupted, not once since each tree sprouted from its seed. Some of these trees have been here since the Roman Empire fell. They were sitting in the California fog when gunpowder was invented, when the Domesday Book was written, when Genghis Khan was rampaging across Asia, when Robert Hooke published *Micrographia* and when the Japanese bombed Pearl Harbor. And not once, in all that time, has the water stopped flowing. The reason we can be sure of this is that the whole mechanism relies on the flow never stopping. There is no way to restart it. But this is very clever plumbing, and the fabulous piece of living architecture that keeps it all going only works because it's just a few nanometres across.

The water travels in the xylem, a system of tiny cellulose pipes that reach through the tree, stretching from the roots to the leaves. This is mostly what 'wood' is, although the innermost wood stops helping with the plumbing as the tree gets bigger. Capillary action, the mechanism that made my towel absorbent, is only strong enough to suck water upwards for a few metres in the tree's plumbing. That's no use for a tall tree. The tree roots can also generate their own pressure to push water up the pipes, but that too is only enough to push the water a few more metres upwards. Most of the work isn't being done by pushing. The water is being pulled. The same system operates in all trees, but the redwoods are the kings of it.

I sit on a fallen tree trunk, just next to one of the giants, and look up. A hundred metres above my head, tiny leaves flutter in the breeze. To photosynthesize, they need sunlight, carbon dioxide and water. The carbon dioxide comes from the air, and it enters the tree through tiny pockets on the underside of each leaf, the stomata. Part of the inner wall of each of those pockets is a network of cellulose fibres, and in between the fibres are water-filled channels. These are the top of the water pipes, after those pipes have branched and branched again, reducing in size each time until they reach the stomata. Here, where the water pipes finally touch the air, each one is approximately 10 nanometres across.[1] The water molecules stick firmly to the cellulose sides of each channel, and the water surface curves down into a nanoscale bowl-shape in between. Sunlight heats up the leaf and the air inside it, and sometimes it gives enough energy to one of those surface water molecules to pull it away from the mob below it. That evaporated water molecule drifts out of the leaf into the air. But now the nanoscale bowl is out of shape – it's too deep. Surface tension is pulling it inwards, pulling the water molecules closer together to reduce the surface area. There are lots of new molecules that could fill the gap, but they are all further back in the channel. So the water in the channel is pulled forwards to replace the lost molecule. And then the water further back in the channel has to shuffle along to replace that, and so on down the tree. Because the channel is so tiny, the surface tension can exert an enormous pull on all the water below it, enough (when you include the contribution of a million other leaves) to pull the entire column of water up the tree. It's a staggering thought. Gravity is pulling the entire tree's worth of water downwards, but the combination of many tiny forces is winning

[1] One nanometre really is tiny – there are a million of them in one millimetre.

the battle.[1] And it's not just a battle against gravity; the upward forces are also defeating the friction from the tube walls as water is squeezed through the tiny channels.

Pushing up from the forest floor around me are the real babies – trees that are just a year old. Their water columns are just beginning to take shape. As the new tree grows, the pipe system stretches but never breaks, and so the top of the water column is always wetting the inside of the stomata. Water is just pulled up towards the air as it keeps growing. The tree can't refill the pipe if it empties, but it can keep the pipe full as it grows. However tall the tree gets, this water column must never be broken. The reason why the tallest redwoods are near the coast is that the coastal fog helps their leaves stay moist.[2] Less water needs to reach the top from the roots, so the system can be slower and the trees taller.

This process of water evaporating from the leaves of trees is called transpiration, and it's happening whenever you look at a tree in the sunlight. These sleepy giant redwoods are actually massive water conduits, sucking it from the forest floor, rerouting some of it for photosynthesis and then letting the rest escape into the sky. It's the same for every tree. Trees are a vital part of the Earth's ecosystems, and they wouldn't be able to climb up into the sky unless they could take water with them. And the beauty of it is that they don't need an engine or an active pump to do that. They just shrink the problem, solve it using the rules

[1] But there's a limit. To increase the tension in the water to pull it up further, the stomata must get smaller. And smaller stomata let in less carbon dioxide, so there is less raw material for photosynthesis. Theory suggests that the tallest a tree could possibly grow is 122–130 metres, because beyond that it wouldn't be able to take in enough carbon dioxide to do any actual growing.
[2] There's also some evidence to suggest that the fog might actually go the other way too – entering the stomata to keep them full of water, not just preventing evaporation.

of the small, and then repeat that process so many million times that it becomes the physics of giants.

The tiny world where surface tension, capillary forces and viscosity dominate over gravity and inertia has always been a part of our everyday life. The mechanisms may be invisible, but the consequences are not. And these days we're not just spectators admiring the elegance and exoticism of what happens down there. We're beginning to be engineers, working within it. There's a word for the rapidly developing field of Lilliputian plumbing, the manipulation and control of fluids flowing through narrow channels: 'microfluidics'. It's not a familiar word to most of us now, but it's going to have a huge impact on our lives in the future, especially when it comes to medicine.

Today, people with diabetes can monitor their blood sugar using a simple electronic device and a test strip. A tiny drop of blood touched to the test strip will immediately whoosh into the absorbent material due to capillary action. Tucked away in the tiny pores of the strip is an enzyme, glucose oxidase, and when this reacts with blood sugar it produces an electrical signal. The hand-held device measures that signal, and *voilà!* – an accurate measure of blood sugar appears on the screen. It's easy to see this as a description of the obvious – paper soaks up a fluid so that it can be measured. So what? But this is just a crude demonstration of the principle. It gets a lot more sophisticated than that.

If you can move a fluid through tiny tubes and filters, gather it in reservoirs, mix it with other chemicals along the way and see the results, you have all the components of a chemistry lab. No need for glass test tubes, hand-held pipettes and microscopes. This is the premise of the growing 'lab-on-a-chip' industry, the development of tiny devices to carry out medical tests. Nobody likes having a whole vial of blood extracted from them, but a single drop isn't too hard to part with. Smaller diagnostic devices are often cheaper to make and easier to distribute.

And you don't even need to make them from fancy modern materials like polymers or semiconductors. Paper might do just fine.

A group of researchers at Harvard, led by Professor George Whitesides, is on the case. They have engineered diagnostic test kits about the size of a postage stamp, made of paper, but containing a maze of water-loving paper channels with waxed water-hating walls. When you touch a drop of blood or urine to the correct part of the paper, capillary action drags it through the main channel, splits it up and reroutes it to lots of different test zones. Each one contains the ingredients to do a different biological test, and each reservoir will change colour depending on the test results.[1] The researchers suggest that someone a long way from a doctor could do the test locally, take a picture of the result with their phone and e-mail it to a distant expert who could interpret it. As ideas go, it's brilliant. Paper is cheap, the device doesn't require power, it's lightweight and a flame is all you need to dispose of it safely. As with all these devices, it's got a lot of checks and balances to face before we know whether the simple-sounding idea can deal with the real world. But it's hard not to be convinced that one way or another, devices like this will be a big part of medicine in the future.

The genius of all this is that when we look at a problem, we may be able to choose to do the engineering on the size scale that makes the problem easiest to solve. It's like being able to choose which laws of physics you want working for you. Small really is beautiful.

[1] These devices go by the catchy name of 'microfluidic paper-based electrochemical devices' or uPADs for short. A non-profit organization called Diagnostics for All has been set up to move this idea into the real world.

4

A moment in time

The march to equilibrium

ON A LAZY SUNDAY at lunchtime, an English pub is the place to be. The innards of these establishments often give the impression of having been grown rather than designed – a cluster of oddly shaped spaces hidden in an ancient oak skeleton. You park yourself at a table positioned between polished brass bedpans and pictures of Georgian prize pigs, and order a proper pub lunch. It always arrives with a bowl of chips and a glass bottle of ketchup, but this combination comes at a price. For decades, these oak beams have witnessed an enduring ritual. The ketchup must be extracted from the bottle, and it won't happen without a fight.

It starts when one optimist picks up the ketchup and just holds it upside down over the bowl of chips. Nothing ever happens, but almost no one skips this step. Ketchup is thick, viscous stuff, and the feeble pull of gravity isn't enough to shift it from the bottle. It's made this way for two reasons. The first is that the viscosity stops the spices sinking downwards if the bottle is abandoned for a while, so you don't ever need to shake it to make sure it's well mixed. But more importantly, people prefer a nice thick coating on each chip, and you can't get that if the ketchup is runny. However, it's not on the chip yet. It's still in the bottle.

After a few seconds, having established that this bottle of ketchup is just as immune to gravity as every other one they've ever encountered, the hopeful chip-eater starts to shake the bottle. The shaking gets gradually more violent, until it's time to try thumping the bottom of the bottle with the free hand. Just as the others at the table are starting to lean back to avoid the fracas, a quarter of the contents of the bottle whooshes out all at once. What's weird is that the ketchup clearly can flow very easily and quickly – the thick blanket of ketchup now covering the bowl (and probably half the table too) provides ample evidence of that. It just doesn't, until it does, and then it flows with considerable enthusiasm. What's going on?

The thing about ketchup is that if you try pushing on it very slowly, it behaves almost like a solid. But once you force it to move quickly, it behaves much more like a liquid and flows very easily. When it's sitting in the bottle or perched on a chip, it's only being pulled weakly by gravity, so it behaves like a solid and stays put. But if you shake it hard enough and start it moving, it behaves like a liquid and moves very quickly. It's all about time. Doing the same thing quickly and slowly gives you completely different results.

Ketchup is mostly sieved tomatoes, jazzed up by vinegar and spices. Left to itself, it's thin and watery, and doesn't do anything interesting. But lurking in the bottle is 0.5 per cent of something else, long molecules made up of a chain of linked sugars. This is xanthan gum: originally grown by bacteria, it's now a very common food additive. When the bottle is standing on the table, these long molecules have surrounded themselves with water and are slightly tangled up with other similar chains. They hold the ketchup in place. As our ketchup enthusiast shakes the bottle harder, these long molecules get slightly untangled, but they re-tangle pretty quickly. As the thumping on the bottom of the bottle shunts the ketchup around more quickly, the tangles keep breaking, and at some point they're pushed out of place more quickly than they're re-tangling. Once this critical point is passed, the solid-like behaviour vanishes and the ketchup is on its way out of the bottle.[1]

---

[1] This behaviour is called 'shear-thinning' and it's also handy for snails, as we'll see shortly.

There is a way around this problem, but considering how much time the British spend eating chips with ketchup on top, it's surprisingly rare to see it. The tactic of turning the bottle upside down and thumping it on the bottom doesn't help much, because the ketchup that is forced to become liquid is all up near where you're thumping. The neck of the bottle is still blocked by thick gloop that isn't going anywhere. The solution is to make the ketchup in the neck liquid, so the thing to do is to hold the bottle at an angle and tap the neck. The amount that will come out is limited, because only the ketchup there is liquid. Surrounding diners will be saved from your elbows (and a potential ketchup spray), and the chips will be saved from drowning.

Time is important in the physical world, because the speed at which things happen matters. If you do something at twice the speed, sometimes you get the same result in half the time. But quite often you get a completely different result. This is pretty useful, and we use it to control our world in all sorts of ways. There's also a lot of time to play with, in the sense that there are a lot of different timescales on which things can happen. Time matters for coffee and pigeons and tall buildings, and the timescale that matters is different for each of them. This isn't just about tweaking the mundane things in our lives for the sake of convenience. It turns out that life is only possible at all because the physical world never really catches up with itself. But let's begin at the beginning. We'll start with a creature that famously never catches up with anything, the mascot for those who are always last.

*

One sunny day in Cambridge, I finally had to admit that I had been defeated by a snail.

It's not traditional to take up gardening during your final year as an undergraduate, but the house I was sharing with three friends had a garden, and the temptation was too much. In the odd spare hours between work and sport that year, I enthusiastically chopped down the huge forest of nettles that had taken over, and discovered buried treasure in the form of rhubarb plants and rose bushes. My Dad laughed at me for planting potatoes ('typical Pole', he said), but they were only part of my new vegetable patch. Most exciting of all, there was a grimy greenhouse, filled with rubble and a grapevine. Seedlings (leeks and beetroot, I think) could grow in shelter before joining the vegetable patch in the spring. In late February I sowed seeds in trays, and waited for new plants to grow.

After a while, it was noticeable that there weren't any seedlings, but there were a lot of snails. I'd arrive with my watering can to find a smug mollusc parked in the middle of each tray, surrounded by bare soil and the occasional green hint of a chewed-off shoot. Not to be defeated, I chucked the snails outside, resowed the seeds, and placed the trays on top of bricks to make it harder for the snails to crawl in. Two weeks later, the nascent seedlings were gone, and there were more snails than ever. I tried a few different approaches, none successful, until I had only one idea left. This time, I took pairs of empty flower pots and balanced upside-down teatrays on top, so they were like giant mushrooms with two stalks. I greased the edges of each one, and put the seedling trays on top of the teatray mushrooms. After replacing the compost, I sowed the last of the seeds, crossed my fingers and went back to studying condensed matter physics.

The seedlings grew undisturbed for about three weeks. And then the inevitable day came when I found one fat, happy snail where the seedlings should have been. I remember standing in the greenhouse, analysing in forensic detail the possible routes that this creature could have taken. There were only two. Option one: it could have crawled up the inside walls of the greenhouse and out across the underside of the roof, and then somehow have dropped off at exactly the right location to land on a seed tray. This seemed unlikely. Option two: it had crawled along the bench and up the flower-pot sides, slimed its way upside down to the outer edge of the teatray, crawled around the edge without falling off, and then trekked along the top of the teatray to the seedlings. In either case, I had to admit that it had probably earned the bounty.[1] How could a snail *do* that? Both cases involved it crawling upside down, while glued to the surface only by its own mucus. If you watch a snail move, it's different from a caterpillar – it doesn't lift itself off the surface as it goes. It's just stuck to its slime, and yet somehow it manages to shift itself along. But that slime is the snail's secret weapon, because it behaves just like ketchup.

If you watch a snail moving, you won't see very much because the outer rim of its foot is just moving at a constant slow speed. Everything on the edges is happening slowly, so the mucus is like the stationary ketchup: thick and gloopy and hard to move. But underneath, in the middle, muscular waves are travelling from the back of the snail to the front. Each wave is pushing forward on the mucus really hard, and it's forcing the mucus to shift very quickly. And just like the ketchup, the mucus is shear-thinning, so if you're shunting it very fast it suddenly flows very easily. The snail is sailing over the top of this liquid mucus on those muscular waves, taking advantage of the lower resistance. It needs the thick slime as well, so that it has some-thing to push against. The only reason why snails (and slugs) can move is that the same mucus can behave either like a solid or like a liquid, depending on how fast they force it to move. The huge advantage of this method is that they don't fall off the

---

[1] Of course, there is a third option: that the snail had been an egg or a juvenile hiding inside the compost. But it was pretty large, and I couldn't imagine it growing so big in such a short time.

underside of things because they never lift themselves away from the surface.

How does the slime manage this trick? It's a gel of very long molecules called glycoproteins, all mixed up together. When it's just sitting still, chemical links form between the chains, so it behaves like a solid. But when you push hard enough, the links suddenly break, and all the long molecules can slither over each other like strands of spaghetti. Let it sit still again, and the links will re-form; and after only a second, you'll have a gel again.

If I had known all this, could I have protected the seedlings? Not by choosing a surface that they couldn't stick to or climb over, it turns out. The mucus can stick to pretty much anything you'd find around the house – including non-stick pan liners. Experiments have shown that snails can even stick to super-hydrophobic surfaces, the ones that water can hardly touch. It's a pretty amazing achievement, but probably one best appreciated by those who don't have precious seedlings to protect.

The same mechanism also explains non-drip paint. When this paint sits still, it's thick and gooey. But when you push on it with a brush, it becomes much less viscous, and it's easy to spread a thin, even layer of it over a wall. As soon as you take the brush away, the paint goes back to being very viscous and so it doesn't run off down the wall before it's dried.

*

Ketchup and snails are small, but this same bit of physics can have serious consequences on a much larger scale. Christchurch in New Zealand was a charming and peaceful city when I visited it in 2002. The land there is made up of sediment, layer upon layer of tiny particles deposited by the Avon river over successive millennia. It's a beautiful location, but the city was sitting on a time bomb. At 12.51 p.m. on 22 February 2011, a magnitude 6.3 earthquake struck just 6 miles away from the city centre. The earthquake itself was bad enough, throwing people into the air and tearing buildings apart. But the sediment that the town was built on was only strong and solid if it was stationary. Just like the ketchup, powerful shaking turned it into a liquid. The small-scale details are a tiny bit different – instead of the bonds between long molecular chains breaking, water sneaks in between sand grains and pushes them apart, allowing them to flow. But the overall physics is the same: when it's agitated quickly, the solid ground starts to flow like a liquid.

A car is a heavy object, so gravity makes it push down hard on the ground it's sitting on. Cars don't sink through the ground because the ground is solid enough to resist that push. But for a few minutes in Christchurch, that general rule was broken. Many cars that day were parked on sandy roadsides, resting on packed soil that hadn't moved for decades. As the earthquake shook the ground, the layers of sand were forced to slide over each other from side to side extremely quickly. If this had happened slowly, the cars would have been safe. But it happened so quickly that water crept in between the sand grains and the sand grains didn't have time to settle back into place before they were forced back in a different direction. So instead of sand resting on sand, the ground was suddenly made of a mixture of sand and water that had no fixed structure. Any car sitting on top of this mixture would sink downwards into the mush as the shaking continued. But as soon as the shaking stopped, it only took a second or so for the sand grains to settle slightly, so that they were supported by other sand grains instead of water. The ground had resolidified, but by now the car was half-buried.

This process was responsible for a lot of the damage in Christchurch. Cars sank into the silt and buildings fell because the ground couldn't hold them up. It's known as 'liquefaction', and it takes something as powerful as an earthquake to move the sediment fast enough to cause it. But if you move soft, sandy ground fast enough, its strength will vanish. This is also why flailing about in quicksand is such a terrible idea. If you fight and struggle, the quicksand becomes liquid-like and you'll just sink in. Move slowly, and you stand a chance of controlling where you are. Time matters. When you change the timescale of what you're doing, you often change the outcome.

We like to say that something was so fast 'it happened in the blink of an eye'. A blink takes about a third of a second, and the average reaction time of a human is around a quarter of a second. That sounds pretty fast, but just think about what has to happen in that time if you're taking a standard reaction test. When light rays hit your retina, specialized detection molecules twist around, and this starts a chain of chemical reactions that cause a small electric current. This signal travels through the optic nerve into the brain, stimulating brain cells to send signals to each other as they work out that this is something that requires a reaction. Then electrical signals travel out to the muscles, slowed down when they are ferried across the gaps between nerve cells by chemical diffusion. Once the order to contract has been received, molecules in the muscle fibre ratchet over each other until your hand hits the button. All that, just for you to do the fastest thing you can possibly do.

Our fabulous complexity comes at the cost of speed. I think of humans as pretty slow beasts, lumbering through the physical world, because so many different stages are involved in every-thing we do. While we plod through all that, many simpler physical systems are just getting on with things, lots of things. But those simple, quick processes are too fast for us to see. You can get a hint of this world if you drop a single drip of milk into your coffee from quite high up. You might just see the drop bounce right back out before falling back into the drink. It's right on the edge of the fastest we can see. My PhD supervisor used to say that if you were quick, you could change your mind about having the milk and catch it on its way out, but I'm pretty sure you'd need the help of something smaller and faster than a human to do it.

The thought of how much we're missing because we're slow is what inspired my PhD. I was fascinated by the idea of a world that could be doing things right in front of my eyes, things that were too small and too fast to see. So I chose a PhD that let me play with high-speed photography, technology that let me see the parts of the world that are normally invisible because they are so fast. But cameras like that are only available to humans. What do you do if you have the same problem, but you're a pigeon?

In 1977, an enterprising scientist named Barrie Frost persuaded a pigeon to walk on a treadmill. This is one of those experiments that would probably win an IgNobel prize these days, as the perfect example of a piece of science that makes you laugh and then makes you think. As the treadmill belt slowly moved backwards, the bird had to walk forward to stay in the same place. The pigeon apparently got the hang of it quite quickly, but something was missing as it plodded along. If you've ever sat in a town square and watched pigeons strut around in search of food, you'll have noticed that their heads bob backwards and forwards as they walk. I've always thought it looks like a really uncomfortable thing to do, and it seems odd to put in all that extra effort. But the pigeon on the treadmill wasn't bobbing its head, and that told Barrie something very important about the bobbing. The bird obviously didn't need to do it in order to walk, so it wasn't anything to do with the physics of locomotion. The head-bobbing was about what it could see. On the treadmill, even though the pigeon was walking, the surroundings stayed in the same place. If the pigeon held its head still, it saw exactly the same view all the time. That made the surroundings nice and easy to see. But when a pigeon is walking on land, the scenery is constantly changing as it goes past. It turns out that these birds can't see 'fast' enough to catch the changing scene. So they're not really bobbing their heads forwards and backwards at all. They thrust their head forward, and then take a step that lets their body catch up, and then thrust their head forwards again. The head stays in the same position throughout the step, so the pigeon has more time to analyse this scene before moving on to the next one. They get one snapshot of their surroundings, and then they jerk their head forwards to get the next snapshot. If you spend a while watching a pigeon, you can convince yourself of this (although it takes a bit of patience, because they are usually quite quick).[1] No one seems to know exactly why some birds are so slow at gathering visual information that they need to bob their heads, and others aren't. But the slower ones can't keep up with their world without breaking it down into a freeze-frame movie.

Our eyes can keep up with our walking pace, but if you need to examine something close to you while you're walking or running, you usually feel an overwhelming urge to stop for a bit to have a good look. Your eyes can't collect information fast enough to get all the detail while you're moving. Humans actually play exactly the same game as the pigeon (without the head-bobbing), and our brain stitches things together so that we'd never know. Our eyes dart rapidly from place to place, adding information to our mental image on each stop. If you look at yourself in a mirror

[1] There's a genuinely funny bit in Frost's paper when he describes what happened when they accidentally set the treadmill to a very low speed. It's not often that I'd quote a scientific paper for comic effect, but in this case it's absolutely justified: 'After completing the filming of a particular bird, the treadmill was inadvertently turned to a very slow setting instead of completely off as intended. After a short time we noticed that the bird's head was slowly and progressively pushed forward until it eventually toppled over. Further observations indicated that toppling, or extreme changes in posture, could also be produced by very slow forward (opposite direction to that eliciting normal walking) treadmill movements. It appeared that the extremely slow (imperceptible to us) speed of the treadmill was not sufficient to induce walking in the bird, but was sufficient to stabilize its head even though this sometimes resulted in loss of equilibrium.'

and look directly at the reflection of one of your eyes and then the other, you will notice that you never see your eyes move, even though someone standing next to you will see them flick from one side to the other. Your brain has stitched your perception of the scene together in such a way that you'd never know there was a jump; but those jumps are happening all the time.

The point is that we're only a tiny bit faster than the pigeon, and this highlights how much there must be that's faster than us. We are used to life at a limited range of timescales – we can follow things that last from about a second to a few years – but that's not all there is. Without science to help us, we are blind to anything happening over a few milliseconds or over a few millennia. We can only perceive our bit in the middle. That's why computers can do so much and part of why they seem so mysterious. They can do what they need to do in tiny amounts of time, so they can get on with it and finish amazingly complex tasks before we perceive any time passing. Computers continue to get faster, but we can't perceive why, because a millionth and a billionth of a second are the same to us: both too fast to notice. But that doesn't mean the distinction isn't significant.

What you see depends on the timescale on which you are looking. To grasp the contrast, let's compare the speedy and the ponderous: a raindrop and a mountain.

A large raindrop takes one second to fall 6 metres, the height of a two-storey building. What happens to it during that second? This raindrop is a jostling cluster of water molecules, each one held firmly in the grip of the group, but constantly shifting its allegiances within that group. A water molecule, as we saw in the last chapter, consists of an oxygen atom accompanied by two hydrogen atoms, one on either side, the trio forming a V-shape. The whole molecule can bend and stretch as it hops through the loose network formed by billions of identical others. In that one second, this molecule may hop 200 billion times. If our molecule reaches the edge of the multitude it will find that there's nothing outside the droplet that can compete with the huge attraction of the masses, so it's always pulled back to the centre. The cartoon raindrop shape is a fiction: raindrops have lots of shapes but none of them have sharp points. Any pointed edges will be rapidly smoothed away, because individual molecules can't resist the pull of the mob. But despite the strength of that pull, the perfect shape is never reached. There is constant readjustment in response to the buffeting of the air. A drop may be squashed flat, but will then pull itself back together, overshoot, become stretched into a rugby ball shape and then back again, 170 times in this one second. The globule is constantly wobbling and re-inventing itself, a battleground between the external forces trying to tear it apart and the fierce pull of the mob keeping it together. Sometimes a raindrop flattens into a pancake, then stretches into a thin umbrella, and then explodes into an army of tiny droplets. All of this happens in less than a second. We can't see any of it, but that droplet has transformed itself a billion times in the blink of an eye. Then the droplet splats down on to bare rock, and the timescales shift.

This rock is granite. It has not moved or changed in human memory. But four hundred million years ago there was a giant volcano in the southern hemisphere, and magma from below squeezed into the gaps in the volcanic rock. Over the following millennia the magma cooled, separating slowly into crystals of different types, and became hard unyielding granite. As more time has passed, the rocky leviathan has been ground down by ice ages, chipped away by plants and ice, polished by rain. While the volcano was wearing away, it was also travelling. Since the giant explosion that finished it, this chunk of continent has been creeping north. On top of it, species and geological eras came and went as the machinery of the planet shunted the ill-fitting jigsaw pieces of its surface together and apart. Today, a tenth of the total lifetime of our planet later, all that is left of the original dramatic volcano is the sorry remains of its exposed guts. We call it Ben Nevis, the highest mountain in the British Isles.

When you and I look at either the mountain or the raindrop, we notice very little change. But that's just because of our own perception of time, not because of what we're looking at.

We live in the middle of the timescales, and sometimes it's hard to take the rest of time seriously. It's not just the difference between *now* and *then*, it's the vertigo you get when you think about what 'now' actually is. It could be a millionth of a second, or a year. Your perspective is completely different when you're looking at incredibly fast events or glacially slow ones. But the difference hasn't got anything to do with how things are changing; it's just a question of how long they take to get there. And where is 'there'? It is equilibrium, a state of balance. Left to itself, nothing will ever shift from this final position because it has no reason to do so. At the end, there are no forces to move anything, because they're all balanced. The physical world, all of it, only ever has one destination: equilibrium.

Imagine a lock gate in a canal. Locks were invented for the most ingenious of reasons: to allow boats on a canal to climb hills. They work because boats can propel themselves forward against water flow, but only if that water flow is really slow. No canal boat can power up a waterfall, but with the help of a lock, a boat can still climb a hill. A lock consists of two sets of gates which form a complete bottleneck in a canal, trapping an isolated pool of water between them. On one side of the lock the water is higher; on the other, it's lower. Anything wanting to go up or down the canal has to go through the lock. Let's say there's a boat waiting at the bottom. The water in between the gates is initially at the same height as the canal at the bottom. The lower gates open, our boat chugs into the lock, and the lower gates close. Now the top gate is opened, just slightly, and water flows into the lock. This is the important bit. When the top gates were closed, the water above the lock had no reason to go anywhere.

It was sitting in the lowest place it could be, in equilibrium. There was nowhere better for it to be, and it would stay put there indefinitely. But as soon as a gap is opened that connects it with the pool of water between the gates, this changes. Suddenly, there's a route to somewhere better. Gravity is always pulling the water downwards, and we've just opened the door for the water to respond to gravity's pull and move itself even further downwards. So it flows in to join the boat, and it keeps filling the lock up until the water height inside it is the same as the water height above the lock. No one had to do anything other than provide the route to a new equilibrium. But now the boat is at the same height as the top part of the canal, and once the gates are fully opened it can chug along on its way upstream, against the very slow canal flow. Behind it, once the gates are closed again, everything is in equilibrium. The water between the locks will stay there indefinitely because it has nowhere better to be. All the forces are balanced. Then at some point a boat will enter the lock from upstream, someone opens the lower gate and the water is allowed to flow out into the downstream canal, where it will continue on its way to a new equilibrium.

The lesson of all this is that you can get a lot done in the world by controlling where the equilibrium position is. Left to themselves, things shuffle around until everything is balanced and then they stay there. The way to get things done is to be in control of where equilibrium is. If you can move the goalposts on demand, you can make sure that things flow in the direction you want them to go in, and only when you say so.

The idea that the physical world will always move towards balance – that hot and cold liquids will mix until everything is the same temperature, or that a balloon will expand until the pressure is equal inside and out – is related to the concept that time only flows one way. The world can't run backwards. Water is never going to flow by itself through a lock from the lower level to the higher level. That means that you can tell which way is forwards by looking for systems moving towards equilibrium. While moving things by brute force will cost you a lot of energy, influencing the speed of the slide to equilibrium often costs very little. It is also often extremely useful.

The Hoover Dam is one of the biggest civil engineering achievements of the last century. Driving towards it from Las Vegas, you weave through a red rocky landscape where it seems impossible that anything large could be hidden. The only clues that there might be something unusual nearby come from occasional glimpses of sparkling blue water completely out of place in the middle of a desert. And then you turn a corner and there it is, all 7.5 million tons of it: a giant concrete bung lodged in the middle of this rugged American landscape.

A hundred years ago, the Colorado river ran unfettered through its narrow canyon. Rain from high up in the Rocky Mountains and the vast plains to the east was funnelled downhill through a series of valleys and out to the Gulf of California. The problem for the farmers and city-dwellers downstream wasn't the amount of water – there was plenty of it – but the timing of its arrival. In the spring, huge floods could wash away fields, but by the autumn only a feeble trickle was left, not enough for the growing population. The water was always going to start in the same mountains and plains, and end up in the same bit of ocean. But what farmers and townspeople alike really needed was to control when it got there,[1] and especially to stop it all arriving at once. And so the bung was built.

---

[1] When I first moved to the American south-west, I couldn't shake off a nagging curiosity about exactly where all the water came from in this dry environment. The book that answered many of my questions (and tells the fascinating story of the battles over water supply in that area) is *Cadillac Desert* by Marc Reisner, and I highly recommend it. California is suffering from a severe drought as I write this, and the tough decisions about how to deal with it cannot be delayed any longer.

A drop of water that's made its way off the Rockies and all the way down through the Grand Canyon now finds itself in Lake Mead, the giant reservoir that has built up behind the dam. It hasn't got anywhere else to go, at least not for a while. The key thing here is that the droplet is held where it is – high up – because it can't go down any further. In 1930, a droplet leaving the Grand Canyon would have trickled 150 metres downwards before it came to rest. But after 1935, when the dam was completed, that same drop could reach this point and still be 150 metres above the valley floor. The amazing thing is that it doesn't take any energy to keep it there, just a carefully placed obstacle to stop it going anywhere else. It's in human-created equilibrium and it's staying put.

Until, of course, humans decide that they want it to go somewhere else. Humans can control the flow through the dam, rationing the water that feeds the rest of the Colorado river. There are no more floods downstream, and the river never stops flowing completely. And there's another benefit. As the marshalled water flows past the dam, the huge pressure that has built up turns turbines that produce hydroelectric power. The consequences of this water shunting are that hundreds of thousands of people can live and work in the arid deserts of the American south-west.

The Hoover Dam was built to control the timing of water flow, but the principle it demonstrates goes far beyond water use. When it comes to harvesting energy, all we are ever really doing is providing a few obstacles to energy that was already on its way from somewhere to somewhere else. The physical world will always move towards equilibrium, but sometimes we can control where the nearest equilibrium is and how quickly something in the world can get there. By controlling that flow, we also control the timing of energy release. Then we make sure that as the energy flows through our artificial obstacles on its way towards equilibrium, it does something useful for us. We don't create energy and we don't destroy it. We just move the goalposts and divert it.

Like many civilizations before ours, we face the problem of limited resources. Fossil fuels are made of up plants that built themselves using energy from the Sun, diverting that energy from its alternative outlet: gentle warmth, which is the equivalent of the bottom of the river when it comes to usefulness. Fossil fuels are the energy equivalent of dams, a form that stores energy in a temporary equilibrium. When we come along, dig them up and provide the right kick, we're choosing the timing of energy release by providing a route to another accessible equilibrium, via a flame and chemical decomposition to carbon dioxide and water. The problem we have is that there are only so many 'upstream' resources in the form of fossil fuels, and in a few human lifetimes we have released energy that took millions of years to accumulate. The fossil-fuel reservoirs are being emptied, and they will not be refilled for millions of years more. Renewable energy, like the hydroelectricity from the Hoover Dam and many others, diverts the waterfall of solar energy that is flowing through our world now. The game facing our civilization remains the same: how do we stop and start the energy flow efficiently, so that we can do what we want without changing our world too much?

Next time you turn on something that is battery-powered, you're choosing the time of energy release from the battery by opening an electrical gate, and guiding the energy through the circuits of the device to help you do something useful. After that, it'll end up as heat, which it would have done anyway. This is what the switches in our world are, all of them. They're the gatekeepers controlling the timing of a flow, and the flow is only ever going one way: towards equilibrium. If we let the flow whoosh through all at once we get one result; if we slow it down, letting it trickle through at times that suit us, we get an entirely different result. Time matters here because it's only ever going in one direction: by choosing when the flow towards equilibrium happens, and the speed of that flow, we give ourselves enormous control over the world. But it's not always the case that things reach equilibrium and then stop. If they're going really fast as they approach the balance point, they may well just keep going and fly right through. This opens the door to a whole new set of phenomena, including some problems.

Mid-afternoon teabreak is an essential part of my working day. But I noticed recently that even acquiring a mug of tea forces me to slow down, and it's not just about the time taken to boil the kettle. My office at University College London is at one end of a long corridor, and the tearoom is at the other end. The journey back to my office, accompanied by a full mug of tea, happens at the slowest pace of my entire day (my normal walking speed at work is somewhere between 'brisk' and 'race pace'). It's not that there's too much tea in the mug; the problem is the sloshing. Every step makes it worse. Any sensible person would accept that slowing down is a reasonable solution. But any physicist would do some experiments first, just to see whether that's the *only* solution. You never know what you might find. And I wasn't going to give in to the obvious without a fight.

If you put water in a mug, sit the mug on a flat surface and give it a bit of a push, the water will start to slosh from side to side. What's happening is that as you shove it, the mug moves but the water initially gets left behind, so it piles up against the side of the mug you've pushed. Then you have a mug that has higher water on one side than the other, so gravity pulls the higher water down, and the water on the other side is pushed upwards. For an instant the surface is flat again, but the water has no reason to stop moving. It just carries on going up the other side. Gravity is tugging on it as it goes, but it takes a while to stop the water completely. By the time it's stopped, the water level is higher on the second side than the first, and then the cycle starts all over again. If the mug is sitting on a flat surface, the sloshing from one side to the other will gradually die away, and equilibrium will be reached. But if you're walking, things are different.

The cycle is where the problem lies. If you try the shoving test with mugs of a few different sizes, you'll see that the sloshing happens in the same way for them all, but it happens more quickly in a narrow mug and more slowly in a wider mug. A mostly full mug always sloshes the same number of times each second, however big the initial push was. But that number depends on the mug, and the thing that matters most is the mug radius.

There's a conflict between the downward force of gravity, which is pulling everything back to equilibrium, and the momentum of the fluid, which is greatest just as it passes through the equilibrium position. In a bigger mug, there's more fluid and it has further to go, so the cycle takes longer to turn around. The special frequency that each mug has is known as its natural frequency, the rate it will slosh at if pushed and then left to get back to equilibrium by itself.

I spent a while playing with the mugs in my office. I have one tiny one with a picture of Newton on it that is only 4 cm across. Water in this one sloshes about five times each second. The biggest one is about 10 cm across, and it sloshes about three times each second. This large mug is old and cheap and ugly and I've never really liked it, but I still have it because sometimes you just need a lot of tea.

When I come out of the tearoom with my full mug and take a couple of brisk steps down the corridor, I start the sloshing. If I want to get back to my office without spilling the tea, I have to prevent this sloshing from growing. This is the crux of the problem. As I walk, I can't help rocking the mug slightly. If the pace of that rocking matches the natural frequency of the sloshing, the sloshing will grow. When you push a child on a swing, you push in a regular rhythm that matches the rate of the swing, and so the swing gets bigger. The same happens with the tea. This is called resonance. The closer the external push is to the natural frequency of the sloshing, the more likely it is that tea will be spilled. The problem for all thirsty humans is that it just so happens that most people walk at a pace that is very close to the natural sloshing frequency of the typical mug. The faster you walk, the closer to it you get. It's almost as if the system was designed to slow me down, but it's just an inconvenient coincidence.

So it turns out there isn't really a satisfactory solution. If I use the tiny mug, it sloshes too fast for my walking pace to make the sloshing worse and the tea won't spill. But I want more than a thimbleful of tea. If I use the larger mug, my brisk walking is very close to its natural frequency, and disaster is just three steps down the corridor. The only solution is to slow down, so that the rocking from the walking is much slower than the sloshing frequency.[1] I feel better for having tried, but the lesson here is that I can't beat the time-dependency of the physics.

Anything that swings – oscillates – will have a natural frequency. It's fixed by the situation, and the relationship between how hard the pull to equilibrium is and how fast things are going when they get there. The child on the swing is just one example, along with a pendulum, a metronome, a rocking chair and a tuning fork. When you're carrying a shopping bag and it seems to be swinging at a rate which doesn't match your steps, that's because it's just swinging at its natural frequency. Big bells have deep notes because their size means that they take a long time to squish and stretch and squish again, so they ring with a low frequency. We get a huge amount of information about the

---

[1] There is actually another solution: start drinking cappuccino. Having a foam layer on top dampens down the oscillations a lot, so foam-covered drinks don't slosh as easily. This is also useful in the pub. Beer snobs may not like too much of a head, but at least it stops them spilling their drink.

size of objects by listening to them, and it's because we can hear how long they take to vibrate.

These special timescales are really important for us, because we can use them to control the world. If we don't want the oscillation to grow, we have to make sure that the system isn't pushed at its natural frequency. That's the game with the tea. But if we want an oscillation to continue without much effort, we choose to nudge it along at its natural frequency. And it's not just people who use this. Dogs do, too.

Inca is poised and ready, focused on the tennis ball like a sprinter waiting for the starting gun. As I lift the plastic arm holding the ball, she tenses, and then the ball sails over her head and she's off, a slim bundle of enthusiasm and seemingly endless energy. Her owner Campbell and I chat while Inca rushes happily across the scrubby grass. She doesn't bring the thrown ball back to us, because she's already got a second tennis ball in her mouth (apparently this is a 'spaniel thing'), but when she reaches it she stands guard until we catch her up and lob the first ball further ahead. After half an hour of non-stop chasing, she finally sits down, tail cheerily swishing the grass, and looks up at us, panting.

I kneel down and stroke her back. All that running around has made her hot. She isn't sweaty because dogs don't sweat, but she still has to get rid of all that excess heat. The panting looks like hard work, presumably using lots of energy and generating even more heat. It seems like a bit of a paradox. Inca is untroubled by my ponderings but quite happy to be stroked, and a strand of saliva drips from her wide-open mouth. After I've been out running, my breathing rate comes down back to normal quite gradually, but when Inca stops panting it happens very suddenly. Big brown eyes look up at me, and I wonder how much longer she needs to recover before it's time for more tennis balls.

By far the most efficient way to lose heat is to evaporate water. That's why we sweat. Turning liquid water into a gas takes a huge amount of energy, and conveniently the gas then floats away, taking that energy with it. Since dogs don't sweat, they don't produce water on their skin that can evaporate, but they have plenty of water in their nasal passages. Panting is all about pushing as much air as possible over the wet insides of their noses, to get rid of heat quickly. As if to demonstrate the point, Inca starts panting again. I reckon she's taking about three breaths each second, which seems like a lot of hard work. But the really clever bit is that it isn't. Her lungs act as an oscillator. This is the most efficient rate for her to breathe at because it's the natural frequency of her lungs. As she breathes in, she's stretching the elastic walls of the lungs, and after a while, the elastic pushes back strongly enough to turn the cycle around. Just as the lungs get back to their unstretched size, she puts in a tiny bit of energy to send them off on the cycle again. The downside is that when she's breathing this fast, she's not really replacing the air deep in her lungs, so she isn't actually taking on board much extra oxygen while all this is going on. That's why she doesn't breathe like this all the time. But just at the moment the need to lose heat trumps her need for oxygen, and by pushing her lungs at exactly the right frequency, she's getting as much air as possible through her nose for as little effort as possible. So the panting is generating a tiny amount of heat compared to what she's losing. She's breathing in through her nose, but she's got her mouth wide open because the dribbling is also cooling her. When the saliva evaporates, that helps lose a bit of heat energy too. The panting stops again, and Inca eyes up the abandoned tennis ball. One inquiring look at Campbell is enough (he's well trained) and the game begins again.

The natural frequency of something depends on its shape and what it's made of, but the biggest factor is its size. This is why smaller dogs pant faster. They've got tiny lungs, which naturally inflate and deflate many more times each second. Panting is a very efficient way of losing heat if you're small. But it gets less efficient as you get bigger, and that may be why larger animals sweat instead (especially hairless ones like us).

Every object has a natural frequency, and often more than one if there are different possible patterns of vibration. As the objects get bigger, those frequencies generally get lower. It can take quite a push to make a really massive object move, but even a building can vibrate, very very slowly. A building can in fact behave a bit like a metronome, a sort of upside-down pendulum – the base is fixed while the top moves. Higher up, the wind is faster than at ground level, and this is enough to give tall narrow buildings the sort of shove that will start them swaying at their natural frequency. If you've ever been in a tall building on a very windy day, you've probably felt this. A single cycle can take a few seconds. It's disconcerting for humans inside, so the architects of these buildings spend a lot of time working out how to reduce the swaying. They can't remove it completely, but they can change the frequency and flexibility to make it less noticeable. If you feel it happening, don't worry – the building will have been designed to bend, and it isn't going to fall.

The wind may be gusty, but it doesn't push in a regular rhythm that could match the building's natural frequency, so there's a limit to how bad the swaying can get. But the jolt of an earthquake sends out ripples in the ground, huge waves travelling out from the epicentre, slowly tipping the earth from side to side. What happens when a tall building meets an earthquake?

On the morning of 19 September 1985, Mexico City started to move. Tectonic plates underneath the edge of the Pacific Ocean, 217 miles away, ground over each other to generate an earthquake of magnitude 8.0 on the Richter scale. In Mexico City, the shaking lasted for three to four minutes, and it tore the city to pieces. It's estimated that ten thousand people lost their lives, and massive damage was inflicted on the city's infrastructure. Recovery took years. The US National Bureau of Standards and the US Geological Survey dispatched a team of four engineers and one seismologist to assess the damage. Their detailed report showed that a horrid coincidence of frequencies was responsible for a lot of the worst damage.

First of all, Mexico City sits on top of lake-bed sediments that fill a hard rock basin. The earthquake monitoring devices showed beautiful regular waves with a single frequency, even though normally earthquake signals are much more complex than that. It turns out that the geology of the lake sediments gave them a natural frequency of oscillation, and so they had amplified any waves that lasted about two seconds. The whole basin had temporarily become a tabletop shaking at almost exactly one single frequency.

The amplification was bad enough. But when they looked at specific damage, the engineers discovered that most of the buildings that had collapsed or were badly damaged were between five and twenty storeys high. Taller or shorter buildings (and there were plenty of both) had survived almost untouched. They worked out that the natural frequency of the shaking closely matched the natural frequency of the mid-sized buildings. With a long-lasting regular push at exactly the right frequency, these buildings had been twanged like tuning forks, and they didn't stand a chance.

These days, controlling the natural frequency of buildings is taken very seriously by architects. Management of shaking is even sometimes celebrated. In the Taipei 101, a 509-metre monster in Taiwan that from 2004 to 2010 was the largest building in the world, the place to visit is the viewing galleries on the 87th–92nd floors. This section of the building is hollow, and suspended inside it is a 660-ton spherical pendulum, painted gold. It's beautiful and weird – and practical. It's there not just as an aesthetic quirk, but to make the building more earthquake-resistant. The technical name for it is a tuned mass damper, and the idea is that when there's an earthquake (a common occurrence in Taiwan), the building and the sphere swing independently. When an earthquake starts, the building sways one way and pulls the spherical pendulum sideways too. But by the time the sphere has moved in that direction, the building has swayed back the other way, and is now tugging the sphere back. So the sphere is always pulling in the opposite direction to the movement of the building, reducing its sway. The sphere can move 1.5 metres in any direction and it reduces the overall oscillation of the whole building by about 40 per cent.[1] The humans inside would be far more comfortable if the building never moved. But earthquakes shove the building out of equilibrium so that it has to move. The architects can't stop that happening, but they can tweak what happens on the return journey. The occupants of the building have no choice but to sit tight as the huge tower sways past the equilibrium position and back again, until the energy is lost and serene stasis is restored.

*

The physical world is always ticking along towards equilibrium. This is a fundamental physical law, known as the Second Law of Thermodynamics. But there's nothing in the rules to say how quickly it has to get there. Every injection of energy kicks things away from equilibrium, moves the goalposts, and the winding down has to start all over again. Life itself can exist because it exploits this system, using it to shunt energy around by controlling the speed of flow towards equilibrium.

Plants still sneak into my life, even though I live in a big city. From my kitchen, I can see bright sunlight falling on the lettuce seedlings, strawberry plants and herbs on the balcony. The light falling on the wooden decking is absorbed by the wood, which heats up, and that heat is eventually dispersed through the air

[1] There are also two smaller pendulums that help with this, just below the main one.

and the building. Equilibrium is reached quite quickly, and nothing very exciting happens along the way. But the sunlight that falls on those coriander leaves is entering a factory. Instead of being converted straight into heat, it's diverted to serve the needs of photosynthesis. The plant uses the light to boot molecules out of equilibrium, and so keeps the energy for itself. By controlling the easiest path back towards equilibrium, the machinery of the plant uses that energy in stages, to make molecules that act as chemical batteries, and then uses those to convert carbon dioxide and water into sugars. It's like a fantastically complex system of canals carrying energy, complete with lock gates, bypass sections, waterfalls and waterwheels, and the flow of energy is controlled by changing the speed at which it passes through each section. Instead of streaming straight to the bottom, the energy is forced to build complex molecules on the way. These aren't in equilibrium, but the plant can store them until it needs their energy, and then it places them somewhere where they can take the next step down towards equilibrium, and then the next step after that. As long as light is falling on to that coriander plant, it's supplying the energy to keep the factory on the hop, continually chasing after equilibrium as the injection of energy moves the goalposts. Eventually, I'll eat the coriander, and that will provide an injection of energy to my system. I'll use that energy to keep my own body from equilibrium, and as long as I keep eating, the system won't be able to keep up. Equilibrium won't be reached. But I choose when to eat, and my body chooses when to use that energy, all by controlling the floodgates.

Considering how common life is on this planet, it's surprising that no one can come up with a single definition of what it is. We know it when we see it, but the living world can usually provide an exception to any simple rule. One definition has to do with maintaining a non-equilibrium situation, and using that situation to build complex molecular factories that can reproduce themselves and evolve. Life is something that can control the speed at which energy flows through its system, manipulating the stream to maintain itself. Nothing that is in equilibrium can be alive. And this means that the concept of disequilibrium is fundamental to two of the great mysteries of our time. How did life start? And is there life anywhere else in the universe?

Scientists currently think that life may have started in deep-sea vents, 3.7 billion years ago. Inside the vents was warm alkaline water. Outside was cooler slightly acid ocean water. As they mixed, at the surface of the vent, equilibrium was reached. It seems that early life may have started by standing in the middle of that path to equilibrium, and acting as a gatekeeper. The flow towards equilibrium was diverted to build the first biological molecules. That first tollgate may then have evolved into a cell membrane, the city wall around each cell which separates inside, where there is life, from outside, where there isn't. The first cell was successful because it could hold back equilibrium, and that was the gateway to the beautiful complexity of our living world. The same is probably true for other worlds.

It seems highly likely that life does exist elsewhere in the universe. There are so many stars, with so many planets, and so many different conditions, that however freaky the conditions needed to form life are, they will have happened in other places. But the chances of that life telling us that it's there by sending us a radio signal are small. Quite apart from anything else, space is so large that by the time any signal reaches us, the civilization that created it would probably be long extinct. However, it may be that the mere existence of life could be broadcasting signals out into the cosmos, completely unintentionally. On the summit of Mauna Kea in Hawaii there is a pair of telescope domes, matching giant white spheres parked next to each other on a ridge. My first thought when I saw them was that they were like giant frog's eyes peering out into the cosmos. This is the Keck Observatory, and it may be these giant eyeballs that see the first hints of life outside our solar system. As alien worlds pass across the front of the distant stars that they orbit, starlight shines through the atmosphere, and those gases leave a fingerprint on the light. The Keck telescopes are starting to pick up those fingerprints, and soon they may be able to detect atmospheres that are not in equilibrium. Too much oxygen to be sustainable, too much methane . . . these could betray the existence of life down on the planet, altering the balance of its world as it strains away from the jaws of equilibrium. We may never know for certain. But that may be the closest we ever come to knowing that there are other organisms out there: the evidence of something controlling the speed of the march to equilibrium, as it builds living complexities that we will never see.

5

Making waves

From water to wifi

WHEN YOU GO TO the beach, it's almost impossible to stand for any length of time with your back to the sea. It feels wrong, both because you're missing out on the grandeur of the sight and also because facing the other way stops you keeping an eye on what the ocean might be up to. And it's oddly reassuring to watch the boundary between sea and land as it constantly renews and remodels itself. When I lived in La Jolla, California, my reward after a long day was to wander down to the ocean, sit on a rock, and watch the waves as the sun went down. Just a hundred metres off shore, the waves were long and low, difficult to see. As they rolled towards the shore they'd get steeper and more obvious until they finally broke on the beach. I could sit and watch the endless supply of new waves for hours.

A wave is something that we all recognize, but they can be hard to describe. The ones at the seashore are processions of ridges, a wiggly shape in the water surface that is travelling from over there to over here. We can measure them by looking at the distance between successive wave peaks and the height of the peaks themselves. A water wave can be as tiny as the ripples you make when you blow on your tea to cool it, or bigger than a ship.

But waves have one quite weird feature, and in La Jolla it was the pelicans that made it obvious. Brown pelicans live all along that coast, and they look so ancient that you wonder whether they've just flown through a wormhole from a few million years ago. They've got ridiculously long beaks that usually stay folded up against their bodies, and small groups of these curious birds are often seen gliding solemnly just above the waves parallel to the coast. Once in a while, they'd plonk themselves down unceremoniously onto the ocean surface. And this was the interesting bit. The waves that the birds were sitting on rolled endlessly towards the shore, but the pelicans didn't go anywhere.

Next time you stand on the shore and watch waves rolling towards you, watch the seabirds sitting on the surface.[1] They'll be parked quite happily, passengers being carried up and down as the waves go past, but they're not going anywhere.[2] What this tells you is that the water isn't going anywhere either. The waves move, but the thing that is 'waving' – the water – doesn't. The wave can't be static; the whole thing only works if the shape is moving. So waves are always moving. They carry energy (because it takes energy to shift the water into the wave shape and back again), but they don't carry 'stuff'. A wave is a regular moving shape that transports energy. I think this is partly why I found sitting on the beach and looking out to sea so therapeutic. I could see how energy was continually carried towards the shore by the waves, and I could see that the water itself never changed.

Waves come in many different types, but there are some basic principles that apply to them all. The sound waves made by a dolphin, the water waves made by a pebble and the light waves from a distant star have a lot in common. And these days, we

[1] One of the unintentional discoveries from my time at sea is that the best way to wind up a bird enthusiast is to casually ask about seagulls. There are gulls (lots of different types), and some of them live in or on or near the sea. But there is no such thing as a seagull. Bird enthusiasts will either spend hours explaining this to you or leave in a huff.
[2] If you get the chance to see them from the side, you'll see that they're actually going round in small circles. The point is that they're not travelling with the wave.

don't just respond to the waves that nature provides for us. We also make our own, very sophisticated, contribution to the flood, and it connects the scattered elements of our civilization. But humans consciously using waves to cement cultural bonds isn't new. This story began centuries ago, in the middle of a gigantic ocean.

A king surfing the ocean waves probably sounds like a snapshot from a particularly weird dream. But 250 years ago in Hawaii every king, queen, chief and chiefess owned a surfboard, and royal prowess at the national sport was a source of considerable pride. Special long narrow 'Olo' boards were reserved for the elite, while the commoners used the shorter and more manoeuvrable 'Alaia' boards. Contests were common, and provided the central drama for many Hawaiian stories and legends.[1] When you live on a stunning tropical island surrounded by deep blue ocean, building a culture around playing in the sea sounds perfectly sensible. But the Hawaiian surf pioneers had something else going for them: the right sort of waves. Their small island nation in the middle of a vast ocean was perfectly placed. Hawaiian geography and physics filtered the complexity of the ocean, and kings and queens surfed on the consequences.

While the Hawaiians were chanting to urge the flat, windless sea to rise into ready-to-surf swell, the ocean thousands of miles away could have looked very different. The winds in massive storms shove on the ocean surface, dumping energy by forcing the water up into waves. But the waves in storms are confused mixtures of short and long waves travelling in different directions, breaking and rebuilding and clashing. Winter storms are common at a latitude of about 45°, so the storms would be to the

---

[1] Other Pacific islanders, most notably the Tahitians, also had surfboards. However, it seems that they only lay or sat on them. The Hawaiians pioneered the idea of standing up on the board, and so 'surfing' as we understand it today.

north of Hawaii in the northern hemisphere winter, and to the south of Hawaii in the southern hemisphere winter. But waves have to travel. Even as the storm winds were dying down, the patch of ruffled ocean would have been expanding out past the edges of the storm and into undisturbed water. Out here, a sorting process could take place. The true nature of the confused mess would be revealed – not jumbled chaos, but a crowd of different wave types all sitting right on top of each other. Water waves that have a longer wavelength (that's the distance between peaks) travel faster than those with a shorter wavelength. So the first waves to escape would be the longest, racing outwards ahead of their shorter cousins. But there is a price to pay as a water wave travels. Energy will gradually be stolen by the surroundings, and the price per mile is higher for the shortest waves. Not only are they losing the race, they're losing their power as well, and it doesn't take too long for them to vanish. Thousands of miles from the storm and days later, all that remains are the longest waves, a smooth regular swell, radiating out across the planet.

So Hawaii's first advantage is being in a spot far enough away from the massive storms to experience them only in the form of that residual smooth, tidy, long-wavelength swell. Its second advantage is that the Pacific Ocean is very deep and the islands' volcanic sides are steep. Waves travel across the ocean surface undisturbed until they suddenly meet a steep slope. Then all the energy that was spread over a huge depth has to become more concentrated in the shallows, so the height of the waves must increase. And very close into shore, the Hawaiians were waiting for the last gasp of these slow monsters, as the waves became so steep that they had to break over the perfect beaches of the islands. And as they broke, the kings and queens were ready with their surfboards.

Water waves are probably the first waves that most people are aware of. Something that a duck can bob about on is easy to imagine and to understand. But waves come in lots of different types, and many of the same principles apply to them all. All waves have a wavelength, a measurable distance between one peak and the next. Because they're moving, all of them also have a frequency, the number of times they go through a cycle (peak to trough and back to peak again) in one second. All waves have a speed, too, but some of them (like the water waves) travel at different speeds depending on their wavelength. The problem with most waves is that we can't see what's doing the waving. Sound waves travel through air, and they're compression waves; instead of a moving shape, what's passed along is a push. The hardest waves to imagine are the most common of all: light waves, which move through electric and magnetic fields. But even though we can't see electricity, we can see the effects of light being a wave all around us.[1]

One of the main reasons that waves are interesting and useful is that the environment they're passing through often changes them. By the time a wave is seen or heard or detected, it's a treasure trove of information because it carries the signature of where it's been. But that signature is only stamped in relatively simple ways. There are three main things that can happen to a wave: it can be reflected, it can be refracted, or it can be absorbed.

*

If you wander past the fish counter at a supermarket and look at what's on offer, what you see is mostly silver. The exceptions to the rule are tropical fish like red mullet and red snapper, and the

[1] Experiments showing that light behaves like a wave were relatively straightforward. It took an extremely clever experiment the size of the Earth's orbit around the Sun to reveal the most counter-intuitive thing about light: there isn't any 'stuff' that's doing the waving. Instead, the waves travel as disturbances in electric and magnetic fields. The test became known as the Michelson–Morley experiment, and it's one of my favourites of all time, because it's simple to understand, it's extremely elegant, and it used our whole planet as a vehicle to test their hypothesis.

bottom-dwelling fish such as sole and flounder. But mostly, you're looking at fish that swim in the open ocean in big schools, like herring, sardines and mackerel. Silver is interesting because it isn't really a colour. It's just our word for something that acts as a trampoline for light, bouncing it back out into the world. All waves can be reflected, and almost all materials reflect some light. What's special about silver is that it sends everything back indiscriminately. Every colour is treated in the same way, no exceptions. Polished metal is really good at this trick, and it's useful because the angle at which the light arrives is the angle at which it leaves. If you take an image of the world and use a mirror to bounce it in a different direction, the relative angles of all those light rays stay the same. It's difficult to polish metal perfectly enough to get a perfect image, and mirrors have been very highly prized in human history. And yet we take silver fish for granted. The fish can't even use metal; in order to be silvery, they've got to build structures that do the same job out of organic molecules. That's complicated, and therefore expensive in evolutionary terms. If you're a herring, why do you bother?

Herring roam the seas in schools, feeding on small shrimp-like creatures and hoping to avoid the big carnivores: dolphins, tuna, cod, whales and sea lions. But the oceans are huge, empty places with nowhere to hide. The only solution is invisibility, or the closest that nature can come to it: camouflage. So should fish be blue, to match the watery background? The problem with that is that the exact hue depends on the time of day and what's in the water, so it changes all the time. But the herring absolutely must look like the water behind them, in order to survive. So they turn themselves into swimming mirrors, because the empty ocean behind them looks exactly like the empty ocean in front of them. They can reflect 90 per cent of all the light that falls on them, similar to a high-quality aluminium mirror. By bouncing light waves back out into the eyes of potential predators, a herring can swim about behind a shield made of light.

Reflection isn't always this perfect. Quite often, only some of the light is reflected by an object. But that's fantastically useful if two objects are sitting next to each other and we want to tell them apart. The one reflecting blue light is my tea mug, and the one reflecting red is my sister's. So reflection matters when a wave hits a surface. But it's not the only thing that can happen when a wave meets a boundary. Refraction can shunt waves about in a more subtle way, altering how they travel.

When a Hawaiian queen stood on a cliff overlooking the coast, watching the surf build, she would have noticed that even though the swell out on the open ocean was approaching from a different direction each day, at the point the water waves reach the shore, they are always parallel with the beach. Waves don't ever come in sideways, whatever direction the coast faces. That's because the speed of water waves depends on the depth of the water, and waves in deeper water will travel faster. Imagine a long, straight beach with swell coming in from a direction that's slightly to the left of straight-on. The part of the wave crest that's on the right, further away from the shore, is in deeper water. So it travels faster, catching up the closer part of the wave, and the whole wave crest turns clockwise as it moves towards the shore, lining up with the beach. By the time the wave breaks, the wave crest is parallel to the shore. So you can change the direction that a wave is travelling in by changing the speed of some parts of the wave crest relative to others. This is called refraction.

It's easy to imagine changing the speed of a water wave, but what about light? Physicists are always talking about 'the speed of light'. It's an unimaginably gigantic speed, and a crucially important fixture in Einstein's most famous legacies: the Theories of Special and General Relativity. The discovery that there is a constant 'speed of light' was controversial and difficult to accept and brilliant. So it feels a bit like spoiling the party to tell you that you have never in your life detected a light wave that was travelling at the speed of light. Even water will slow light down, and you can confirm this for yourself with a coin and a mug.

Put the coin flat on the bottom of the mug so that it's touching the side closest to you. Now bend down until the edge of the mug just hides the coin from you. Light travels in straight lines, and at this point there is no straight line that can get from the coin to your eyes. Now, without moving your head or the mug, fill the mug up with water. The coin will appear. It hasn't moved, but the light from it changed direction as it left the water and now it can reach your eyes. It's an indirect demonstration that water slows down light. As the light meets the air, it speeds up again and so the wave is bent through an angle as it crosses the boundary. We call this refraction. And it's not just water that does this; everything light passes through slows it down, but by different amounts. The 'speed of light' means its speed in a vacuum, when light is travelling through nothingness. Water slows light down to 75 per cent of that speed, glass to 66 per cent, and light in diamond is dawdling along at 41 per cent of its maximum speed. The more it's slowed down, the bigger the bend at the boundary with the air. This is why diamonds are so much more sparkly than most gems – they slow light down much more than the others.[1] And that bending is the only reason that you can actually see glass, water or diamonds. The material itself is transparent, so we don't see it directly. What we see is that something is messing about with light coming from behind it, and we interpret that something as a transparent object.

It's nice that we can see diamonds (and will come as a relief to anyone who has shelled out for one), but refraction isn't just about aesthetics. Refraction gives us lenses. And lenses opened

---

[1] Like many materials, diamond slows down different colours of light – different wavelengths – by different amounts. So part of the sparkle comes from the diamond separating out the colours, as well as bouncing them back at you.

the doors to a huge chunk of science: microscopy to discover germs and the cells that we're made of, telescopes to explore the cosmos, and cameras to record the details permanently. If light waves always travelled at the speed of light, we would have none of those things. We live in a bath of light waves, and those waves are constantly being reflected and refracted, slowed down and sped up as they travel. Just like the chaos of the stormy ocean surface, overlapping light waves of different sizes are travelling in every possible direction around us. But by selecting and refracting, keeping some waves out and slowing others down, our eyes marshal a tiny fraction of that light so that we can make sense of it. The Hawaiian queen standing on the cliff was watching water waves by using light waves, and the same physics applies to both.

That's all fine if some waves have arrived for you to see after being reflected or refracted. But what if they never reach you at all?

One of life's little oddities is that if you give a child some crayons and tell them to draw water coming out of a tap, the water in their picture is blue. But no one has ever seen blue water coming out of a tap. Tap water has no colour (if yours has, I suggest you seek advice from a plumber). If you did see blue water coming out of a tap, you certainly wouldn't drink it. But the water in pictures is always blue.

On satellite pictures of the Earth, the oceans are definitely blue. It's not because of the salt – there are ponds of salt-free meltwater on top of glaciers, and they are also a stunningly deep, spectacular blue. They almost look like someone has filled pockets in the ice with blue food colouring. But where water is trickling over the ice to join the rest of the meltwater, it has no colour. What matters for the colour isn't what's in the water, but how much water you have.

Light waves hitting the water surface are either reflected back up into the sky or pass through and travel down into the depths.

But sometimes, a tiny particle or even the water itself acts as an obstacle, sending the wave off in a new direction. This redirection may happen to the same light wave enough times that it eventually makes its way back out to the air. And on that long journey, the water has filtered the light. The light waves coming from the Sun are a mixture of lots of different wavelengths, all the colours of the rainbow. But the water can absorb light, and it absorbs some colours more than others. The first to go is the red light – a few metres of water is enough to get rid of most of that. And then the yellows and greens follow after a few tens of metres. But blue light is hardly absorbed at all – it can travel for huge distances. And so by the time the light is on its way out of the ocean, most of what's left is blue. The reason tap water is colourless is that there isn't enough of it to make a difference. Tap water does have a colour, the same colour as all the other water in the world. But that colour is so faint that you need a huge amount of water all together to actually see the effect that the water is having on the waves going through it.[1] When you do see it, it's spectacular, and bright blue crayon really is the right choice. But you'd never learn that from a tap.

So as waves travel, they can be absorbed by whatever they're passing through. It's a very slow process of attrition, sneaking away wave energy tiny bit by tiny bit. The amount that's lost depends on what type of wave it is and also its wavelength. All this variability means there's a huge richness in what waves are doing and what they can tell us. We can see and hear some of the

---

[1] It would be interesting to see the colour chosen by children to draw water in a culture that doesn't have this habit. I think that we identify water as blue because we know about the oceans, and we have aerial photography and very clean swimming pools. But few cultures had those things until recently. Are there enough hints that they would unconsciously colour it blue? Or is that entirely a learned habit?

contrasts in one of my favourite atmospheric phenomena: the thunderstorm.

A thunderstorm is a magnificent spectacle, a dramatic reminder that air is far more than an invisible filler for the sky. Our atmosphere is host to vast quantities of water and energy, and usually these hefty commodities are shunted around slowly and peacefully. The thundercloud, the mighty cumulonimbus, develops in order to rebalance the atmosphere when peaceful shunting is no longer enough. The system starts when buoyant, warm, moist air near the ground shoves upwards into the cooler air above, taking huge amounts of energy with it. In the centre of the vast cloud, hot, humid air rises rapidly, churning the atmosphere above it and liberating huge raindrops. Most dramatic of all, the churning causes electrical charges to be separated and redistributed to different parts of the clouds. The charges accumulate until nearby clouds or the Earth itself are stabbed by giant pulses of electrical current, carrying the excess electrical charge away. Each lightning bolt lasts for less than a millisecond, but the thunder echoes across the landscape for far longer. I love thunder and lightning, both for the theatrical spectacle and for the glimpse it gives us into the atmospheric engine. Thunderstorms produce such unlikely opposites: the sharp, shocking flash of lightning contrasting with the deep, drawn-out rumble of thunder. But both are beautiful examples of the versatility of waves.

The lightning bolt is temporary. The electrical connection is a superheated tube of atmosphere, stretching from the thundercloud to the Earth or perhaps to another cloud. It's a corridor full of molecules that have been blown apart by the energy rushing past them. For a brief instant, the temperature in that tube may reach 50,000°C, and so it blazes blue-white. A giant pulse of light waves whooshes outward from the tube, filling the landscape, but they rush away at such an enormous speed that they're gone in an instant. As the superheated tube carrying the electrical current heats up, it expands sideways, thumping into the air around it. This gigantic pressure pulse ripples outward through the air, following the light, but much more slowly. These are sound waves, and this is the thunder. We know that lightning bolts exist because they make both light and sound waves.

The most important thing about a wave is that it's a way of letting energy move, but without also having to move air, water or 'stuff' of any kind. This means that waves can billow through our world very easily, disturbing things enough to be interesting and useful, but not so much that they're shoving our world about and causing major disruption. A lightning strike liberates a lot of energy, and light and sound waves can carry some of that energy out into the rest of the world, sharing it out. Even though the air doesn't go anywhere overall as the sound ripples past, huge amounts of energy are transferred onwards. Light and sound are different types of wave, but the same basic principles apply to both. For example, both light and sound can be changed by the environment that they pass through. In the case of thunder, we can directly hear what's happening to the waves.

My favourite place to be is about a mile from the lightning strike. Once the flash has signalled that the sound is on its way, I like to imagine that giant pressure ripple spreading out towards me. As I look out across the landscape, I can see right through the ripple, but it takes a few seconds to reach me with the first whipcrack of thunder. These sound waves are travelling at about 340 metres every second or 767 mph, which means they're taking 4.7 seconds to cover a mile. That sharp crack is similar to the original sound made as the lightning bolt expanded right at the ground. But here's what makes the sound of thunder so distinctive: what I hear just after the initial crack is the sound from slightly higher up the lightning bolt. It started as the same sound, but it took longer to reach me because it had to travel a sloping, and therefore longer, path. And then as the thunder rumbles on, I'm hearing the sound from higher and higher up that same lightning bolt. If it takes five seconds for the first crack to reach me, it'll take two more seconds before the sound from one mile up hits me, and another four seconds before the sound from two miles up arrives. All these sound waves started off more or less the same, just in different places. And that means that as I listen, I can hear how the atmosphere is changing these waves. As time goes on, the only difference is that they've travelled further. So the highest-pitched sounds, that first sharp crack, disappear very quickly as the high-frequency waves are absorbed by the atmosphere, but the lower-frequency waves rumble on. As time goes on, and the waves have travelled further and further, the overall pitch gets lower and lower, because the highest notes are consumed by the air, but the lowest notes just keep going. If you're far enough away, the air takes it all and the sound never reaches you. But the lightning has a greater reach – these light waves are different and they don't depend on the air for assistance as they travel. They don't get absorbed by air as easily, but they can be altered in other ways as they whoosh through the world.

In a sense, waves are very simple. Once they've been made, they are always on their way to somewhere else. And whether they're sound waves or ocean waves or light waves, they can be reflected or refracted or absorbed by their environment. We live our lives in the middle of this complex flood of waves, sensing the patterns in those that give us clues about our surroundings. Our eyes and ears tune into the vibrations all around us, and those vibrations carry two very important commodities: energy and information.

*

On a grim, grey, cold winter day, toast is the perfect comfort food. The only problem is that the gratification is not instant. I usually put the kettle on for tea, then put bread in the toaster, and then pace the kitchen impatiently while I wait for my treat to be ready. After I've washed a mug or two and tidied the work surface, I often find myself staring into the toaster, checking on what it's up to. The nice thing about toasters is that you can see they're up to something, because the heating elements glow red. They're not only heating up the air that touches them, they're radiating light energy too. And this glow is a built-in thermometer. You can tell how hot the element is just from its colour. This bright red tells me that the innards of my toaster have reached 1,000°C. That's horrifically hot – enough to melt aluminium or silver. But if it's glowing that bright cherry red, then 1,000°C is how hot it is. It's a rule that comes from the way our universe works. Everything that is this temperature will glow the same colour of red, and other colours indicate other temperatures. If you look into a coal fire and see the innermost coals glowing bright yellow, you know that they are around 2,700°C. Something that is white-hot is 4,000°C or above. But when you think about it, that's odd. Why should colour have anything to do with temperature?

While I'm staring into the toaster, I'm watching energy transform from heat to light. One of the most elegant things about the way the universe works is that anything that has a temperature above absolute zero is constantly converting some of its energy to light waves. And light must travel, so the energy whizzes out into the surroundings. The red-hot heating element is converting some of its energy into red light waves, at the long-wavelength end of the rainbow. But most of the energy it's emitting has even longer wavelengths than that, and we call these waves infrared. Infrared is just like the light we can see, except that each wave is longer. We can only detect it indirectly, by feeling the warmth where it's been absorbed. Even though we can't see them, infrared waves are essential for a toaster – they are what heats the toast up.

Hot objects send out more light at some wavelengths than others. At any temperature, there's a peak wavelength which accounts for most of the light, and the radiated light dies away on either side of that peak. The toaster is sending out a big bulge in the infrared, and the tail of the bulge is visible red. So I see red. I can't see the light that's heating my toast, but I can see the tail of longer wavelengths.

If I had some kind of super-toaster that could get even hotter, perhaps to 2,500°C, the heating elements would look yellow. That's because the hotter object would send out light with shorter wavelengths, so the visible tail would include more of the rainbow: red, orange, yellow, and a little bit of green. When we see both red and green light together, we interpret that as yellow. Only something that has this temperature would send out this exact segment of the rainbow. And if the temperature increased even more – if I had a hyper-toaster that could get to 4,000°C – the light sent out would include the whole rainbow, all the way to blue. And when we see all the rainbow colours at once, we see white. So something that is white-hot is actually sending out a rainbow, but all the colours are mixed up. The disadvantage of the hyper-toaster is that it would melt pretty much whatever you made it out of. But it would brown your toast very quickly. And possibly your kitchen as well.

So a toaster is just a way of making waves. The red light waves that you see are just some of the waves that it has made because of its temperature. The infrared waves that you can't see heat up your toast. This is why toast only browns at the surface in a toaster; it's only the bits that the light touches that can absorb infrared and heat up. The reason I'm quite happy to stare at the toaster while I'm waiting is that I'm imagining all the light it's giving out that I can't see. I know it's there, because the red glow is a giveaway.

But of course, there's a catch. The problem with this method of generating light waves is that you always get the same set of waves together. There's no way to choose some of them but not others. An orange-hot coal and molten steel and anything else that's 1,500°C must emit the same collection of colours all together. So you can measure the temperature of something by its colour, when it's hot enough for you to see the colours. The surface temperature of the Sun is about 5,500°C – that's why it gives out white light. In fact, this is the only reason we can see stars in the night sky; they're so hot that light must pour out from their surface and across the universe, light with a specific colour that gives their temperature away.

And we – you and I – we also have a colour because of our temperature. It's not a colour that we can see, but it's visible to special cameras adapted for the right sort of infrared. We're much cooler than the toaster, but we're still glowing. We emit light waves with wavelengths that are mostly 10–20 times longer than visible light. Each of us is a lightbulb in the infrared, just because of our body temperature. And so are dogs and cats and kangaroos and hippos – all warm-blooded mammals. Anything and everything that is above absolute zero (the scarily cold temperature of –273°C) is a lightbulb like this, with the colour crossing from the infrared to even longer wavelengths (the microwave range) as the temperatures get colder.

So we live our lives bathed in waves, and not just the ones we can see, the ones that might catch our eye if we look in the right direction. The Sun, our own bodies, the world around us, and also the technology we create are constantly making light waves. And the same goes for sound waves – high notes, low notes, the ultrasound that bats use to hunt and the infrasound that elephants use to follow the weather. The amazing thing is that all of these waves can be travelling through the same room, and none of them will interfere with any of the others. The sound waves are the same whether a room is completely dark or full of disco lights. The light waves aren't affected by piano concertos or screaming babies. All of this is what we tap into when we open our eyes and use our ears. We're just siphoning off some of the useful bits from the flood, selecting the waves that send us the most useful information.

But which ones do you choose? The answer will be different for the newest self-driving cars and for an animal that needs to survive in a forest. There's a huge richness of information out there, and you can pick and choose which of the waves will help you most. That is why blue whales and bottle-nosed dolphins can hardly hear each other, and also why neither of them gives a hoot about the colour of your wetsuit.

*

The Gulf of California stretches along the western coast of Mexico, a narrow ocean haven 700 miles long that opens into the Pacific at its southern end. The blue water of the channel is protected by dark, raw mountain peaks that poke into the sky from both shores. Marine species migrate vast distances across the oceans to feed and to rest here. Bobbing about in a small boat in the middle of the channel, a fisherman can appreciate the peace. And what peace means is that the flood of waves that bathes that fisherman is low-key and relatively uncomplicated. Light streams from the Sun during daytime, reflected only by blue water and burnished rock. The lapping waves and the creaking of the boat send out the only sound waves. A lone dolphin leaps out of the water, briefly a part of this calm world, and then splashes back down into a completely different world and one that certainly isn't calm. Down below is the loud, bustling hubbub of an ecosystem at work and play.

The dolphin sends out a high-pitched whistle as it dives downward, communicating to the rest of the pod following on behind. And as the pod catches up, the water is filled with clicks, short sharp waves sent out from the forehead of each dolphin that bounce off the surroundings. Those that make it back to the first dolphin are transmitted through its jawbone to its ear, so each animal builds up a picture in sound of what's nearby. The whistles and squeaks and clicks make this sound like a busy street.

These are the sound waves of a community on the move. After spending a while at the surface, breathing and playing, the pod turns downwards towards the deepest, darkest blue, on a mission: the hunt. The light waves that were so common above the surface are much less common down here. Light waves are absorbed by water very quickly, so information from light is scarce. The dolphins have eyes that can cope both above and below water, but the measure of light's usefulness to them is in how that eye has evolved. They have no ability to distinguish colour at all – why would you need it when there's hardly any variety in the colour of your world? Their world is blue, but they will never know that. A dolphin can't see the colour blue, so their watery world looks black. But they can see the bright glints of passing silvery fish, so they can see what they need.

The ocean surface is like an Alice-in-Wonderland mirror, separating two worlds but easy to step through. Waves tend to bounce off the interface, so sound from the air stays in the air, and sound from the ocean stays in the ocean. In air, light travels very easily, and sound travels reasonably well. In the ocean, light waves are absorbed very quickly, but sound waves pass through quickly and efficiently. If you want to learn about your environment in the ocean, you need to detect sound waves. Light waves are often little use unless you're looking at something very close to you and near the surface.

But there's more to the world of sound down here. Dolphins use very high-pitched sounds, some with wavelengths ten times shorter than anything we can hear. These short wavelengths mean that their echolocation mechanism can pick up on even tiny details of the shape of what's in front of them. But high-pitched sounds don't travel very far, so the noisy pod of dolphins can't be heard from the other side of the channel. On top of the dolphin chatter are other sounds that travel much further. There's the deep hum of a distant ship, the tinkling of bubbles from surface splashes, the quiet popcorn-like crackling of snapping shrimp, and then a deep groaning noise, so low that the dolphins can't hear it. The groan is repeated. Ten miles away, a blue whale is calling and the sound is echoing up the channel. The whale doesn't use echolocation, so it doesn't need a high-pitched wave. But it needs the sound to travel a long way, and that means using a low pitch (a long wavelength). A sound wave with a long wavelength can travel for huge distances, and the baleen whales – blue, fin and minke whales, among others – need to communicate over huge distances. The whales can't hear the dolphin clicks and the dolphins can't hear the whale song. But the water carries it all, a vast flood of information for whichever creatures choose to tune in.

So the ocean has its own flood of light waves and sound waves, but in a completely different way from the air. Sound is king down there, and whales and dolphins are colour-blind because the details of the light waves don't matter to them.

There are some similarities between the atmosphere and the ocean, though. Just as the longest wavelengths of sound travel furthest underwater, the longest light wavelengths travel furthest in the air. Just over a century ago, humans too learned to communicate over thousands of miles. Because we live in air, we don't do it using sound waves. Our long-distance communication uses light waves. When light waves have wavelengths that long, we call them radio waves. And the most important early use of this technology was to send information across oceans. If her crew had really taken in the information carried by these new communication systems, the *Titanic* might never have sunk.

Just after midnight on 15 April 1912, circular pulses of radio waves were rippling outward from a handful of spots in the North Atlantic Ocean. The patterns started and stopped sporadically, and each one faded as the ripple travelled outward from its source. Some of the ripples reached the other spots that were broadcasting, and these were relayed on. The strongest ripples of all came from a spot 400 miles south of Newfoundland in

Canada, where Jack Phillips was using one of the most powerful marine radio transmitters in service to beg for help. The gigantic RMS *Titanic*, the largest ship in the world, was sinking. Jack was up on the boat deck, at the top of the ship, sending short electric pulses up to the antenna which was strung between the funnels. The oscillations in the aerial wire sent crude bursts of radio waves out from the ship, and the radio operators of other ships could decode the pattern and understand the message.

Radio only works because waves like this don't travel in a single direction, but ripple outwards in all directions. You don't need to know the exact position of the person who's listening, and many different people can listen to the same waves. The pulses that the *Titanic* sent out could be detected by the *Carpathia*, the *Baltic*, the *Olympic* and several other ships within a few hundred miles. The information transmitted may have been limited and the means clumsy, but for the first time in human history it was possible to have a conversation across an ocean. The arrival of radio technology changed shipping for ever. Twenty years earlier, the *Titanic* would have disappeared beneath the waves alone and it would have taken a week or so to work out that it had gone. The first transatlantic radio signal had been sent only ten years previously. But that night, via the waves rippling out through the dark, nearby ships were connected to the tragedy as it happened. The staccato pulses weren't random. The ripples came in patterns, and each pattern conveyed a message sent by one human, broadcast out across the vast distances of the ocean at the speed of light. It represented a huge revolution in human communication. This was the roar that signalled the real beginning of the age of radio.

One of the reasons why the demise of the *Titanic* is so famous is that it happened on the cusp of this new age. It showed the enormous potential of radio waves – the RMS *Carpathia* did arrive two hours after *Titanic* sank, in time to save many lives. But it also showed that the radio system of the time was really too crude to be useful. Messages were slow to send, and some of the warnings of icebergs that the *Titanic* had received were lost in the flood of trivial or more general messages. More importantly, using crude bursts of waves meant that signals easily got confused. Who was speaking and who was listening? Messages might not be heard in their entirety, or they might not be heard at all. To use waves to send information, you need to alter them in some way, so that the receiver can see a pattern. But all these ships had was 'on' and 'off' – a burst of radio waves or nothing. There was only one channel, and everyone had to share.

Radio waves weren't the only waves zooming over the ocean that night. The *Titanic* sent up distress flares and the nearby *Californian* tried to communicate with her by using Morse lamps, sending out flashes of visible light. But the radio waves could travel far further, because of a convenient quirk of the atmosphere. An upper atmospheric layer (called the ionosphere) acts as a partial mirror for radio waves. So the radio signals from the *Titanic* weren't just sweeping outward over the surface of the ocean; they were bouncing up into the atmosphere and then back down again. This is why radio waves can travel across oceans, even though the curvature of the Earth means that there's no line of sight between sender and receiver. Reflecting waves can travel around a planet, because the reflections help them get around the curved surface. There's no equivalent mirror in the sky for visible light.

Jack Phillips continued to fill the night sky with pulses of radio waves, broadcasting the ship's position to anyone who was listening until water flooded into the wireless room. He didn't survive, but long-distance communication by radio waves meant that 706 others out of the 2,223 on board did. And they lived to see a world that went from total radio silence to a cacophony of communication via these invisible waves. Almost nowhere on Earth is untouched by them now, and human civilization is interconnected as never before.

Light waves rule our world. They are the vehicle that delivers to us the tiny fraction of solar leftovers that power our planet. They connect us to the rest of the universe. But in the past century, our civilization started to develop a new relationship with the suite of all possible light waves, the electromagnetic spectrum. Where once we were passive consumers, grateful for the energy and information that accidentally came our way, now we are prolific producers and users of light waves. Our sophistication at manipulating light has opened the doors to colossal skill at monitoring our world, the ability to broadcast information almost instantly to almost every living human, and being able to talk, right now, to any individual on the planet in possession of a mobile phone.

But you can only make sense of the flood of waves if you have some way to separate out the many messages being sent. Fortunately, the waves themselves provided the answer, and you don't need any specialist kit to see it for yourself.

The Great Smokey Mountains in Tennessee are spectacular, a huge stretch of valleys and peaks blanketed by deep green forest. The serenity and untouched feel of the forest were particularly striking because to get there we had to drive through Dolly Parton's home town. I knew of the great country singer, of course, but I wasn't prepared for the sight of Dollywood, a huge theme park celebrating Tennessee, country music, fairground rides and of course Dolly herself. And that's just the epicentre. Pink cowboy hats, highly decorated guitars and an all-pervasive background of country music seep into the surrounding towns, along with big blonde hair, vintage denim jackets and a big Southern welcome. Bourbon after dinner seemed to be a cultural imperative, although secretly I'd have preferred a cowboy hat. But it all changed when we got up into the mountains the following day. Crowds flocked in with deckchairs and drinks coolers, and quietly set themselves up to watch the forest. Anything other than complete darkness would spoil the show, so all lights were extinguished, and torches and phones were forbidden. As dusk fell, the dance of the fireflies began. The forest sparkled with the flashing of millions of tiny broadcasting insects. We were there to make a science documentary, and we had just one evening to capture the whole event. The problem with filming something like this is that you need to move about and to be able to see where you're going. We had been told that if we really had to, we could use red lights, as apparently they won't disturb the fireflies as much as white lights. So we crept around the forest, dimly glowing red. By about 1 a.m., the fireflies had mostly stopped, and we were preparing to film the last segment. While the director and cameraman were setting up lights, I parked myself in a pitch-black clearing with my red head-torch, huddled under a piece of blackout cloth because of the cold, and scribbled notes about what I was going to say. When the others were ready, I went to join them, and opened my notebook for a last reminder of the ideas I'd had. But under the director's white head-lamp, I couldn't read my notes. On the page were two sets of scribbles, one in red pen and one in blue, right on top of each other. It was impossible to read either one.

If you wanted to demonstrate how entirely separate different wavelengths are from one another, it would be hard to come up with a better example than this. I realized that I must have written on that page in red ink earlier in the day. Under white light, it's easy to see red ink on white paper. But under a red headlight, red ink is invisible. The white paper reflected the red light back into my eyes. And the red ink also reflected red light back into my eyes. With my red head-torch, the page looked empty because the red light was bouncing back off all of it in exactly the same way. So I wrote new notes on the same page in blue ink. I could see the blue ink because it doesn't reflect red light, so there was a contrast between ink and paper. If I had looked at the page with a blue head-light, I would have been able to see the red ink but not the blue. Just as if I was turning a radio dial, I could have chosen what to read by choosing the colour of illumination I was using. Red light has a longer wavelength than blue light. By selecting the wavelength to pay attention to, I was choosing the information I'd get.

In fact, this is exactly like tuning into a radio station. Most of the ways that we use to detect light (and other sorts of waves) will detect only a very narrow range of wavelengths. If a wave with a different wavelength goes past, we have no way of knowing that it's there. My notebook made it obvious that this is true for the visible colours, but it's just as true for the invisible colours. The world around us is absolutely flooded with different light waves, and they all just sit on top of each other like notes in different-coloured ink. They don't interact with each other or change the other colours that are present. Each one is entirely independent. You can choose to detect very long-wavelength radio waves, and listen to a radio station. Or you can press the button on a remote control that sends out infrared signals that can only be seen by your television. Or you can write in red ink on a page. Or you can wait for your phone to see what wifi networks are available – each network is effectively being broadcast in a different colour, but these colours have microwave wavelengths. This cacophony of information is there all the time, each wavelength just sitting on top of all the others. And it's only if you look for information in the right way that you'd ever know it was there. We paint our picture of the world in a very narrow range of wavelengths, the visible colours of the rainbow. But these visible colours aren't affected in any way by all the other colours out there.

The fact that waves with different wavelengths don't affect each other is really useful. We can pluck out the interesting ones and be conveniently deaf to the rest. Each different wavelength is affected by the world around it in a different way. The world is sorting and filtering the waves, depending on their wavelengths. This is why, although I was brought up near grey, cloudy, rainy Manchester, where seeing the night sky was a rare treat, I lived only 14 miles from the biggest telescope in the UK. The Lovell telescope at Jodrell Bank is a huge radio telescope with a dish 76 metres in diameter. And even on the greyest Manchester days, when rainclouds stack up kilometres thick, this telescope has a perfect view of the sky. For visible light, with a wavelength less than a millionth of a metre, entering a cloud is like entering a giant pinball machine. The light gets bounced and diverted, and is eventually absorbed completely. But the massive radio waves, exactly the same except that their wavelength is about 5 cm, sail straight through all those minuscule obstacles, completely unaffected. Next time you're in Manchester in the rain, bear that in mind. Maybe it will provide some small comfort to think that astronomers can still see the majesty of the cosmos, even if you can't even see the tops of the trees.[1] Or maybe it won't.

Earth is only habitable because different wavelengths of light interact differently with the things they touch. Energy streams out from the hot Sun as a broad symphony of light waves, and our rocky planet intercepts a tiny fraction of the torrent. The energy carried by that tiny fraction is what keeps us warm. But if that was all there was to it, the Earth's average surface

[1] Although astronomers haven't always believed that it's the majesty of the cosmos they're looking at. In 1964, Robert Wilson and Arno Penzias detected waves from the sky at microwave wavelengths that shouldn't have been there. They spent a long time trying to work out which bit of the sky or their telescope was messing up the measurement, sure that something was generating extra microwave light. They also cleared out some nesting pigeons from the telescope, along with their droppings (euphemistically described as 'white dielectric material' in the paper they wrote). The unwanted background light persisted. It eventually turned out to be the signature of the Big Bang, some of the most ancient light in the universe. There is something special about an experiment that has to be very careful to distinguish between the after-effects of pigeon poo and the after-effects of the formation of the universe.

temperature would be a frigid −18°C, rather than its current comfortable 14°C. What saves us from being permanently frozen is the Earth's 'greenhouse' effect. The way it works has to do with different wavelengths of light interacting with the atmosphere in different ways.

Imagine the view from a hillside on one of those cartoon-like days when the sky is mostly blue but there are a few puffy white clouds meandering along to add a bit of variety. If you're looking out over flatter land, you can see green trees, grass and dark earth. Sunlight illuminates the scene, apart from the shadows left by the clouds. But what's reaching the ground in front of you is different from what left the incandescent Sun. The atmosphere has absorbed the long infrared wavelengths, and most of the shorter ultraviolet wavelengths, but the visible light has sailed through unaffected. The atmosphere has already selected the waves that reach the ground. It just so happens that they're the ones we can see. At the visible wavelengths, the sky behaves as an 'atmospheric window', letting everything through. There's another window for radio waves (that's why radio telescopes can see the cosmos), but most of the other waves are blocked by the air.

The darker the land you can see, the more of those visible waves it's absorbing. And the absorbed energy eventually ends up as heat. If you touch dark ground on a sunny day, you'll feel that heat. The rest is reflected upwards, back out through the atmospheric window. If any aliens are out there looking at us, that's what they'll see us with.

But now the ground has heated up. And just like the toaster heating element, it must give away light energy because of its temperature. It's relatively cool, so we can't see the glow. But in the longer-wavelength infrared, the warm ground is a lightbulb. And this is where the greenhouse effect comes into play. Most of the atmosphere will just let these infrared waves through. But some gases – water, carbon dioxide, methane and ozone – punch above their weight. Even though they make up only a fraction of the total atmosphere, they absorb the infrared waves very strongly. They're known as the greenhouse gases. As you look out across the landscape, you can see the visible light leaving the surface, but you can't see the infrared. If you could, you'd see that it faded away as it got further from the ground. The atmosphere is absorbing the infrared waves as they travel upwards. It won't be long before those molecules will give up their new energy, and send it out again as more infrared waves. But here's the important bit. When the new waves are sent out, they'll be sent in all directions equally. Only some will travel upwards and out of the atmosphere. Some will travel back downwards, and be reabsorbed by the ground. So some of the travelling energy is trapped in the atmosphere. That extra little bit of heating is what keeps our planet warmer than it should be, allowing liquid water to exist. A new balance has to be established; ultimately, the same amount of energy must both arrive and leave, otherwise we'd continually be getting hotter. So the Earth heats up until it can give away enough infrared waves to balance the books.

This is the 'greenhouse effect'.[1] Most of it is natural – there's lots of water and carbon dioxide in our atmosphere, and everything is in balance when the average surface temperature is 14°C. But as fossil fuels burn, humans are adding carbon dioxide to the atmosphere, so that more of the infrared energy travelling upwards is trapped. This shifts the balance. So the planet will heat up until a new balance is achieved. The amounts of carbon dioxide involved are very small: CO_2 made up 313 parts per million of the atmosphere in 1960, and 400 parts per million in 2013. Compared with all the other molecules up there, it's a tiny increase. But these molecules select certain waves to absorb. Methane will absorb even more infrared than carbon dioxide. So these gases matter. The greenhouse effect is what made our

[1] Which has very little to do with how a real greenhouse does anything.

planet habitable, but it also has the potential to change the temperature significantly. It's all happening with waves that we can't see directly. But we can measure the consequences already.

There are all sorts of waves rippling around our world – giant radio waves, minuscule visible light waves, ocean waves, ponderous deep sound waves emitted by whales underwater and the high-frequency echo-sounding beacons sent out by bats. Each type is zooming through and past the others, but has no effect on them. But we have one more question to answer. What happens when a wave meets another of exactly its own type? The answer is beautiful if you are holding an iridescent pearl, but something to be avoided if you're trying to hold a mobile phone conversation.

Pinctada maxima can be found parked on the seabed, just a few metres below the surface of a turquoise sea near Tahiti and other South Pacific islands. When it's feeding, the two halves of its shell part slightly and it sucks in sea water, gallons each day. The mollusc inside the shell quietly filters out any valuable specks of food and then expels the cleaned water to rejoin the ocean. You could swim right over the top of it and never notice – the outside of the shell is coarse and unremarkable, mottled in beige and brown. These vacuum cleaners of the ocean look the part: functional and unglamorous. The inside of an oyster was never meant to be seen. And yet Cleopatra, Marie Antoinette, Marilyn Monroe and Elizabeth Taylor were all proud owners of what happened when an oyster's innards made the best of a bad job: pearls. *Pinctada maxima* is the South Pacific pearl oyster.

Occasionally, an irritant finds its way into the wrong bit of the oyster. Since it has no way of expelling the intruder, the oyster coats it in something harmless, the same stuff that it coats the inside of its shell with. It's the mollusc version of sweeping something under the carpet, except that it makes the carpet to fit rather than using one that's already there. The coating is made of tiny flat platelets cemented together with organic glue and stacked up on top of each other. Once it has begun the coating process, the oyster just keeps going. It was recently discovered that the pearl turns as it forms, going round perhaps once every five hours. The tides and seasons come and go, sharks and manta rays and turtles pass by overhead, and the oyster quietly sits there, filtering the ocean as the growing pearl pirouettes slowly in the dark.

Serenity reigns for years, until our oyster has a terribly bad day and is hoicked out of the ocean by a human and prised open. As sunlight hits the pearl for the first time, waves of light bounce off its shiny white surface. But they don't just bounce off the platelets at the top; some make it through to the next few layers and bounce off those instead, or maybe bounce a few times inside the layers before they make their way out. So now we have a situation where there's a single type of wave – let's consider just the green light from the Sun – and it's overlapping with other waves of exactly the same type. The waves still don't affect each other, but they do add up. Sometimes, the green light wave that bounces off the top surface lines up exactly with the green light wave that has bounced off the next surface down. The peaks and the troughs of the wave shape match perfectly. So they keep going out into the world together, a strengthened green wave. But perhaps the red light arriving from the same angle, and bouncing off the layers in exactly the same way, doesn't line up as perfectly. The peaks from one red wave line up with the troughs of the other red wave. Add them together, and there's nothing left to travel in that direction.

These platelet layers are the reason that an obscure filter-feeding mollusc from the South Pacific can make something that's sought after by the most glamorous individuals in human society. The layers are so thin and so tiny that they're just the right size to affect how the light waves line up. The important role they play is to shuffle light about a bit so that waves of the same type overlap. The waves add up (a physicist would say that they interfere with each other), and the result is coloured patterns. From some angles, the reflected light waves reinforce themselves and so we see shimmers of pink and green from the shiny white surface. At other angles, it might be blue that lines up, or no colour at all. As the pearls are turned in the sunlight, we see the flashes that come from the added-up waves. This is what we call iridescence – a mysterious-looking shininess that is highly prized by humans because it's so rare and so beautiful. What's happening is that the pearls are creating an irregular pattern of light waves, and as you move past them, you see different parts of the pattern. But it looks to us almost as though the pearls are glowing, and we love it. More recently, humans have learned to engineer the world on this scale for themselves. But even these days, we still mostly get the oysters to do the hard work for us.

Pearls show what happens when waves of the same type overlap. Sometimes the crests and the troughs line up and add together, making a stronger wave travelling in a specific direction. Sometimes they cancel out, so there's no wave in that direction at all. A new wave pattern is going to result whenever there's anything for waves to reflect from, or when there's more than one source of waves (think of the overlapping ripples from two identical pebbles dropped into a pond next to each other).

But this raises some questions. What happens when other sorts of identical waves overlap? What about mobile phones? We've all seen clusters of people standing close together, all having phone conversations with different people, but using identical phone models. They are connected to the world via waves, the same types of wave as hundreds or thousands of other people in the same city. Radio communication as the *Titanic* went down was hampered because twenty ships in the whole of the North Atlantic were all using the same technology and the same type of wave to send out signals. But today you could have a hundred people in a single building all having separate conversations on identical mobile phones at the same time. How have we managed to organize this cacophony of waves to make that possible?

Imagine looking down on a busy city. A man walking down the street pulls a phone from his pocket, taps at the touchscreen, and holds the phone to his ear. Now, add a superpower to your sight, the ability to see radio waves of different wavelengths as different colours. Green waves ripple outward in all directions from the man's phone, brightest and strongest at the phone itself, dimming as they travel away. There's a mobile phone base station 100 metres away, and it detects the green waves and decodes the message – identifying the number he wants to contact. Then the base station sends out its own signal back to the man's phone, another green ripple, but the colour of this new signal is fractionally different from the original green. This is the first trick of modern telecommunications. Whereas the *Titanic* could only send a signal that was a mixture of lots of different wavelengths, our technology is now incredibly precise about which wavelengths are sent and received. The wavelength of the original signal from the phone was 34.067 cm, and the wavelength used to send the return signal was 34.059 cm. The phone and the base station can listen and talk on channels with wavelengths that are different by only a tiny fraction of 1 per cent. Our eyes can't distinguish colour with anywhere near the same precision. But like the red and blue ink on my white paper, those waves are separate and don't interfere with each other. As the man walks down the street, the green waves rippling out from his phone carry a pattern, the message that is being relayed on. A woman across the street is also talking on the phone, using a fractionally different wavelength again. But the base station can distinguish between the two. This is why the government sells bandwidth as a range; if your phone network is using that range, you're free to make the differences between channels as tiny as you like, as long as the hardware can separate them out.

So as we look down on this section of the city, we see lots of bright spots, as phones send out their signals. The signals are bouncing off buildings and slowly being absorbed by the surroundings, but most of them reach a base station before they get too weak.

As the man we were watching walks down the street, away from the base station, we start to see new colours. The streets ahead of him are full of red radio splotches, all centred on the next base station, which is sending out many shades of red to the phones around it. As the strong green signal from the first base station fades, the man's phone detects the new frequencies and starts communicating with the new base station. He has no idea he's reaching the edge of the 'green' section, but as he does so his phone switches wavelengths so that it's now sending out shades of red. These aren't picked up by the original green base station, but they are relayed on by the new red one. If he keeps walking, he might walk into areas where the radio waves look like yellow or blue to us, with our superhero radio vision. No two patches of the same colour touch; but if he walks even further, he might walk into a new green area. This is the second trick of our mobile phone networks. By keeping the signal strength very low, we make sure that the signals can only reach the nearest base station. That means that a little way further on, you can have a new station, using the same green frequencies. But the signals from the two green stations are too weak ever to reach each other, so there's no problem with interference. Information flows to and from the centre of each cell (that's what the patch around each base station is called),[1] but doesn't interfere with the information from other cells. It doesn't matter that everyone is talking at once, because they're all talking using slightly different waves. And the technology can separate out all these conversations by

[1] That's why mobile phones are called cellphones in US English – the network is cellular.

tuning its receivers with incredible precision. If your phone sends signals at a wavelength that's wrong by a tiny fraction, the message will never get through. But the incredible precision of modern technology means that the tiniest subtleties are enough to tell the waves apart.

And this is what we walk around in every day. Zooming past our heads are overlapping ripples from phones, wifi networks, radio stations, the Sun, heaters and remote controls. And those are just the light waves. On top of that is the sound: the deep rumbles of the Earth, jazz music, dog whistles and the ultra-sound being used to clean the instruments in a local dental surgery. And then the ripples on the cup of tea as we blow on it to cool it, ocean waves, and the undulations of the surface of the Earth itself from the occasional earthquake. And more. We're filling our world with more waves all the time, as we use them to detect and connect the details of our lives. But they all behave in basically the same way. They all have a wavelength. They can all be reflected and refracted and absorbed. Once you under-stand the basics of waves, the trick of sending energy and information without sending stuff, you've got a huge grasp on one of the major tools of our civilization.

In 2002, I was working in New Zealand at a horse-trekking centre near Christchurch. One evening, the phone rang and, to my astonishment, it was for me. The phone had a cordless hand-set, so I could take it outside and sit on the hillside, looking out through the dusk over the New Zealand countryside. It was Nana. She had decided to call me (I'd been away from the UK for about six months by that time, and I hadn't spoken to my family at all), and so she pressed the right numbers on her phone and there I was, on the other end. As her familiar Lancashire accent asked about the food and the horses and the work, I was com-pletely distracted by the weirdness of the situation. I was on the other side of a gigantic planet, as far from my family as it was possible to be while still on Earth (12,742 km in a straight line, and 20,000 km as a very enthusiastic crow flies), and here was Nana on the phone. Just . . . talking. Chatting. But there was a whole planet in the way. I've never quite got over how disconcerting those ten minutes were. These days, our planet is connected by waves. We all talk to each other, all the time, via waves that we can't see. It's such a gigantic achievement, and so fundamentally odd. The work of inventors like Marconi and events like the sinking of the *Titanic* pointed the way towards the world of today where we take these connections for granted. I feel very grateful that I was born just early enough to experience the astonishment that this particular achievement deserves. Our eyes can't detect these waves, and it's always hard to appreciate the invisible. But next time you make a phone call, give it a thought. A wave is really a very simple thing. But if you're clever about how you use it, it can shrink the world.

6

Why don't ducks get cold feet?

The dance of the atom

SALT IS OFTEN CONSIDERED a mundane commodity, stored away in cupboards, never the centre of attention. But if you look a little bit more closely at a handful of grains of salt, especially in bright light, you'll notice that it's surprisingly sparkly. And it gets better as you get closer. Peer at it through a magnifying glass, and you'll see that the grains aren't randomly shaped, or lumpy, or rough. Each one is a beautiful little cube with very flat sides, perhaps half a millimetre across. This is why it sparkles; light is reflecting off those flat faces as though they were tiny mirrors, and different salt grains are glinting at you as you turn the pile in the light. The boring stuff in the salt cellar is made up of minuscule sculptures, each with the same precise shape. Salt manufacturers don't do this deliberately – it's just how salt forms. And it gives us a clue about what 'stuff' is made of.

Salt is sodium chloride, and it's made of equal numbers of sodium and chloride ions.[1] You can think of them as balls of different sizes – the chloride has almost twice the diameter of the sodium. When salt is forming, each one of its components has a

---

[1] An ion is just an atom that's either given away some electrons or gained some extra ones. Here, the sodium atom has given the chlorine atom one electron, so the sodium becomes a positive ion and the chloride becomes a negative ion. It sounds perverse, but now that they've got opposite charge, they'll attract each other.

fixed place in a very specific structure. Like eggs in a stack of giant egg-boxes, the chloride ions assemble in rows and columns, so that they sit on a square grid. The smaller sodium ions fit into the spaces in between, so that each little box of eight chloride ions has a sodium ion in the middle. A salt crystal is just a giant grid like this, a cube that's a million or so atoms long on each side. When the salt crystals grow, they tend to grow a new layer across an entire flat face before they start on the next layer, so the cube keeps its flat sides as it grows. It's atomic-scale filing, each component stacked up perfectly in its place. And the flat sides of each cube can reflect light like a mirror.

We can't see the individual atoms, but we can see the pattern of their structure because the whole salt crystal is just that same pattern repeated again and again. Salt is very simple, and a bigger salt grain is just more of the same. The flat faces that make the salt sparkle are there because individual atoms have to sit in specific places on a rigid lattice.

Sugar also sparkles, but when you look more closely at sugar crystals (especially the larger ones, like those in granulated sugar), you'll see something even more beautiful. These crystals are six-sided pillars with pointy ends. Each sugar molecule is made up of forty-five different atoms, but those atoms are held together in a fixed way that is the same in each individual molecule. One sugar molecule is a brick in a crystalline sculpture, even though it's a brick with quite a complicated shape. Like the simpler salt crystals, these too stack up on top of each other in a regular lattice, and there's only one pattern for them to follow. Once again, we can't see the atoms, but we can see the pattern because the whole crystal is just a giant stack, a skyscraper of molecules. Since the six-sided pillars have flat sides that can act as mirrors, sugar sparkles just like salt.

Flour and rice and ground spices don't sparkle, because they have a much more complicated structure – they're made from the tiny living factories that we call cells. The only reason sugar and salt crystals have perfectly flat sides is that they have such a simple structure: just rows and columns of atoms slotted into specific positions. And that perfect repetition is only possible because at the bottom of it all there are billions of tiny identical building blocks: atoms. The sparkling is a reminder of their existence every time you put a spoonful of sugar in your tea.

Even though we can't see the atoms themselves, we can see the consequences of what's happening down there in the world of the tiny. The goings-on at the bottom of the size scale directly affect what we can do at the largest scales in our society. But first, you have to believe that atoms exist.

These days, we take the existence of atoms for granted. The concept of everything being built of minuscule balls of matter is relatively simple and it makes sense to us because we've all grown up with it. But you only have to go back as far as 1900 to find serious debate in the scientific community about whether atoms were there at all. Photography, telephones and radio had arrived to herald a new technological age, but there was still no agreement on what 'stuff' was actually made of. To a lot of scientists, atoms seemed like a reasonable idea. For example, chemists had discovered that different elements seemed to react in fixed ratios, which made perfect sense if you needed one atom of one sort plus two atoms of another sort to make a single molecule. But the sceptics held out. How could you be sure about whether something so tiny was really there?

Many decades later, a quote was attributed to the scientist and science-fiction writer Isaac Asimov that expresses perfectly the most common path of a scientific discovery: 'The most exciting phrase to hear in science, the one that heralds new discoveries, is not Eureka! (I found it!) but rather, "Hmm . . . that's funny . . .".' The final confirmation of the existence of atoms is a perfect example of science taking that route, but it was nearly eighty years in coming. The clock started ticking in 1827, when the botanist Robert Brown was looking through a microscope at pollen grains suspended in water. Tiny particles were breaking off the pollen, and they were pretty much the smallest things that could be seen by an optical microscope, then or now. Robert Brown noticed that even when the water was perfectly still, those tiny particles were jiggling about. At first, he assumed that it was because the particles were alive, but he later observed the same thing with non-living matter. It was weird, and he didn't have any explanation for it. But he wrote about it, and over the following decades many other people saw the same thing. The weird jiggling became known as 'Brownian motion'. It never stopped, and it was only the tiniest particles that jiggled. Various people proposed explanations, but no one really got to the bottom of the mystery.

In 1905, the world's most famous Swiss patent clerk published a paper based on his PhD dissertation. Einstein is best known for studies on the nature of time and space, and the Theories of Special and General Relativity. But his PhD topic was the statistical molecular theory of liquids, and in papers in 1905 and 1908 he laid out a rigorous mathematical explanation of Brownian motion. Suppose, he said, that the liquid is made of many molecules, and those molecules are bouncing off each other continually. He painted a picture of the liquid as a dynamic, disorganized substance, with molecules hitting each other and speeding up and slowing down and changing direction with each bump. Then, what happens to a larger particle, one that's much bigger than the molecules? It gets bumped from lots of different directions. But because the bumps are random, sometimes the particle gets hit more on one side than the other. So it moves sideways a bit. And then it gets randomly bumped upwards more than downwards. It moves a little because of that. So the jiggling of the bigger particle is just the consequence of its being hit by many thousands of much smaller molecules. Robert Brown couldn't see the molecules, but he could see the bigger particles. The jiggling that Einstein predicted matched what

Brown had seen. And that could only be the case if the liquid really was made up of molecules bumping into each other. So individual lumps of matter – atoms – must exist. Even better, one of Einstein's equations predicted how big the atoms would have to be to cause the jiggling that was seen. And then, in 1908, Jean Perrin carried out even more detailed experiments that fitted Einstein's theory. The last doubters had no alternative but to be convinced by the new evidence. The world was made of many tiny atoms, these atoms constantly jiggled about, and at last everyone could move on. Those two discoveries went hand in hand. The constant vibration of the atoms wasn't incidental, either. It turns out to explain some of the most fundamental physical laws about the way the world works.

One of the biggest consequences of the new understanding of atoms and molecules was that phenomena like Brownian motion had to be explained using statistics. There wasn't any point tracking every individual atom, calculating exactly what happened when it hit another one, and keeping track of each of the billions of atoms in a single drop of liquid. Instead, you worked out the statistics of what would happen, given lots of random collisions. On any given day, you couldn't say that the Brownian particle would go exactly 1 millimetre to the left. But you could say that if you did the experiment lots of times, it would end up 1 millimetre away from its starting point *on average*. You could calculate that average very accurately, but an average was all you were going to get. It meant that physics was a bit messier than it had been in 1850. But it could explain an awful lot. Once you knew about atoms, even everyday things like sodden clothes looked much more interesting.

The first programme I ever presented for the BBC was about the Earth's atmosphere and the patterns of weather around the globe. And so I got to spend three days in the largest and most famous weather event on the planet: the Indian monsoon. The monsoon is a yearly change in the wind patterns around India, and between June and September each year the reversed winds bring rain. Lots and lots of rain. We were there to talk about where all that water was coming from.

We stayed in tiny wooden huts on a very quiet beach in Kerala, right down on the southern tip of India. The first filming day was long and varied – monsoon weather is very changeable, which is frustrating if you need a couple of hours with the same weather conditions to film one particular section. Hot sunshine was followed by an hour of very intense rain, and then strong winds and then back to the hot sunshine. But it was warm all day, and I never mind being rained on if I'm not cold. Being cold really isn't fun. Every time it rained, I got drenched, and then I'd have to work out how to make my clothes look slightly drier when the sun came out. The problem with being the one on camera is that you're the only person who has to wear the same clothes all the time. So I found a sheltered warm sunny nook where things would dry out a bit, and it felt like I spent hours changing into and out of clothes of various degrees of dampness, trying to match the sogginess of my clothes to the current weather conditions. At about 7 p.m., the heavens opened (again), I got drenched (again) and since the sun was going down, we decided to call it a day.

I wrung out my top and shorts as well as I could, used a towel to dry them as much as I could (which got them to 'damp'), and then hung them up and went to get dinner. There they stayed until six the next morning, when it was time to get up and start work. But when I picked up my shorts, they weren't just damp – they were wetter than the night before. And not only were they wet, they were also very cold, because the temperature had dropped overnight. Yuck! But I didn't have any duplicates, so I had to put them on, and walk along the beach trying to look soulful in the sunrise without visibly shivering.

In a gas, the molecules generally aren't attracting each other, which is why they can spread out to fill whatever container you put them into. In a liquid, things are a bit different. The game of bumper cars is still going on, but the molecules are much closer together – so close that they're touching almost all the time. In air at room temperature, the average distance between any pair of gas molecules is about ten times the length of a single molecule. But in a liquid, the molecules are right next to each other. They're still jiggling about as they bounce into the molecules next to them, and they can move past each other quite easily, but they're moving more slowly than the gas molecules. Because they're slower and closer to each other, molecules in a liquid feel the attraction of other molecules near them. That's why liquids form droplets. Temperature is all about the amount of movement energy that the molecules have. In a cool liquid droplet, the molecules aren't moving much and so they stick together. If you heat the droplet up, the average speed of all the molecules will increase, and some will happen to end up with quite a bit more energy than average.

For a molecule to escape from the liquid, it needs enough energy to escape the attraction of the others. This is evaporation, and it happens at the moment a molecule has acquired just enough energy to escape from a liquid and float off by itself to join a gas. My wet clothes were full of liquid water, molecules sluggishly moving around each other but without the energy to escape.

For three days in that monsoon, I tried everything I could to dry my clothes off. Drying clothes generally means putting them in a situation that will give the liquid water molecules enough energy to escape, so that they just drift off elsewhere. During bursts of hot sunshine, the liquid water absorbed the sun's energy and the water molecules would slowly escape. But when it was cloudy, I was fighting a losing battle. The problem was that there was just too much water in the air. The air blowing towards the beach from the ocean was full of it. As the sun shone on the hot ocean, it warmed the surface layer. The water molecules in the ocean are also playing bumper cars, and the hotter the water gets, the faster they move, on average. As the ocean surface heated up, more molecules happened to end up with enough speed to escape. These molecules drifted up into the atmosphere to become gas instead of liquid. So the warm, moist air that arrived at the beach was already full of escaped water molecules. They were now playing bumper cars with the other molecules in the air.

When I got rained on, the warmth from my body was heating up my clothes, giving some of the water molecules I was carrying around enough energy to escape into the air. That was making the clothes drier. But the kicker was that there were so many water molecules in the air that they were bumping into my clothes and sticking. When that happened, they would just sink into the liquid mob, making my clothes wetter. The reason my clothes never dried was that the number of water molecules evaporating from them into the air was exactly balanced by the number of water molecules that were condensing on to them from the air. This is what 100 per cent humidity means: that every molecule that evaporates is replaced by another one condensing. If the humidity is lower than 100 per cent, more molecules will leave the liquid than arrive. The bigger that difference is, the faster things dry.

At night, it got worse. As the air cooled, all the molecules slowed down. So even more of them slowed down enough to stick to my top and shorts, and the clothes got even wetter. The point at which more molecules are condensing than evaporating is called the dew point, and the liquid drops that form are dew. Occasional molecules will still have enough energy to leave the liquid and join a gas. But their numbers are insignificant compared to the molecules that are coming the other way. If I had been able to heat up the clothes, I would have increased the number of molecules evaporating, perhaps enough to tip the balance back so that the clothes got drier. As it was, I was stuck with the wet, and so was the rest of India.

The point is that there's always an exchange going on. That statistical way of looking at a sea of molecules is important because the molecules aren't all doing the same thing. At exactly the same time, in exactly the same place, some molecules will be evaporating and some will be condensing. What we see just depends on the balance between those two possibilities.

There are times when it's really helpful that each molecule in a mob is behaving differently. For example, when sweat evaporates, it's only the molecules with the most energy that escape. The consequence is that the average speed of the ones left behind decreases. That's why sweating cools you down; the escaping molecules take lots of energy away with them.

Clothes generally dry pretty slowly. It's a peaceful process. Once in a while, a particularly energetic water molecule finds itself at the water surface with enough energy to escape, and off it drifts. But it doesn't have to be like that. And violent evaporation can be very useful, especially when you're cooking. It turns out that frying food, generally classed as a 'dry' cooking method, owes an awful lot to water.

My favourite fried food is halloumi cheese, something I've always thought of as the vegetarian's answer to bacon. It all starts with oil heating in a heavy pan, while I chop up rubbery strips of cheese. The oil silently takes in enough heat to raise its temperature to about 180°C, and if I couldn't feel the heat nearby, I'd never know anything was happening. But as soon as I drop in the first strips of cheese, the peace is smashed by loud crackling and sizzling. As the cheese touches the hot oil, its surface layer is heated up to almost the oil temperature in a fraction of a second. The water molecules at the surface of the cheese suddenly have loads of extra energy, far more than they need to escape the liquid and float off as a gas. And so they burst apart from each other, producing a series of mini gas explosions as the molecules in the liquid are liberated. These bubbles of gas are what I can see at the surface of the cheese, and this is where the noise is coming from. But the bubbles have an important role to play. As long as gaseous water is streaming out from the cheese, the oil can't get in. It can barely touch the surface, only just enough to pass on heat energy. This is why frying food at too low a temperature makes it greasy and soggy; the bubbles don't form quickly enough to keep the oil at bay. As the cheese cooks, some heat is transferred into the bulk of the cheese, heating it up. The outer edges give away lots of water, because it's too hot for liquid water to remain there. This is why the outer surface becomes crispy – it's dried out. The browning comes from chemical reactions that happen as the proteins and sugars in the cheese get heated up. But the sudden transition from liquid water to gas is at the heart of how frying works. And frying food has to involve sizzling – if you're doing it right, there's no way to avoid it.

*

The transition from gas to liquid and back again is happening all the time around us. But we don't see the transition from liquid to solid and back nearly as often. For most metals and plastics, melting happens a long way above everyday temperatures. For smaller molecules like oxygen, methane and alcohol, melting happens at fantastically low temperatures, the sort of temperatures that require very specialized freezers. Water is an unusual molecule, since it both melts and evaporates at temperatures that occur around us fairly regularly. But when we think of frozen water, we think most often of the North and South Poles of the Earth. They're cold, white, and forever associated with the great expeditions of the twentieth century that took humans into some of the most inhospitable environments on the planet. Freezing water caused them a lot of problems. But sometimes it also offered unusual solutions.

The transition from gas to liquid is all about molecules getting close enough to each other to touch, while still moving freely enough to flow over each other. The transition from liquid to solid is about the moment those molecules get locked into place. The freezing of water is the most common example of this, but water freezes like almost nothing else. There's nowhere this weirdness is more visible than the frozen north – the Arctic Ocean.

If you travel to the northernmost part of Norway, stand on the coast and look still further north, you see the sea. During the ice-free summer months, the 24-hour daylight nourishes vast mobile forests of tiny ocean plants, a seasonal smorgasbord that attracts fish, whales and seals. Then, towards the end of the summer, the light starts to disappear. The surface water temperature, which only reached 6°C even at the height of the summer, starts to drop. The water molecules, slipping and sliding over each other, slow down. The water is so salty here that it can get down to −1.8°C and stay liquid; but one clear dark night, the ice starts to form. Perhaps a flake of ice is blown on to the water, and if the slowest water molecules bump into it, they will stick. But they can't stick just anywhere. Each new molecule rests at a fixed place relative to the others, and the jumble of bustling molecules is replaced by a crystal, in which well-ordered water molecules are marshalled into a hexagonal lattice. And as the temperature drops further, the ice crystal grows.

The utterly weird thing about water crystals is that the rigorously aligned molecules take up more space now than when they were dashing around in the warm. With almost any other substance, parking molecules on a regular grid would make them sit closer together than when they're allowed to roam free. But water's not like that. Our growing crystal is less dense than the water around it, and so it floats. Water expands as it freezes. If it didn't, the newly frozen ice would sink, and the polar oceans would look very different. But as it is, the temperature drops further, the freezing ice expands, and the ocean grows itself a coat of solid white water.

There are lots of things in the frozen Arctic to get excited about: polar bears and ice and the Northern Lights. But there's one particular piece of Arctic history that I absolutely love, a story that's all about the peculiarities of ice freezing, and of working with nature rather than against it. It's about a bulbous, stout little ship which survived one of the most extraordinary voyages in the history of polar exploration. She's called the *Fram*.

Explorers in the late 1800s were drawn to the North Pole. It wasn't that far away from western civilization. The northern parts of Canada, Greenland, Norway and Russia had all been visited and at least roughly mapped. But the North Pole itself was a big mystery. Was it land? Sea? No one had ever reached the pole, so no one knew. The voyage defeated explorers again and again because the sea ice grew and shrank and shifted. As weather conditions changed, sea ice could pile up on top of itself, making ridges and ice quakes. The grip of this ice could grind ships to pieces. The USS *Jeanette* suffered a typical fate in 1881, becoming trapped in sea ice for months just off the northern coast of Siberia. As the weather cooled, and molecules of sea water locked on to the bottom of the ice lattice at the sea surface, the expanding ice gripped the hull. After months of ice growing and shrinking, squeezing then releasing the ship, the USS *Jeanette* succumbed and was crushed. Explorers who made it off their vessels on to solid ice faced different perils: the ice could melt and open up huge canals, impassable except with a boat. From any of the countries around the Arctic Circle it was hundreds of miles to the pole, and the shifting ice was a formidable obstacle.

Three years after she sank, unmistakable wreckage from the USS *Jeanette* washed up near Greenland. It was an astonishing find because the wreckage had crossed the entire Arctic, right from one side to the other. Oceanographers wondered whether there was a current that left the coast of Siberia, travelled across the North Pole, and carried on to Greenland. And a young

Norwegian scientist called Fridtjof Nansen had a wild idea. If he could make a ship that would withstand the ice, he could take it to Siberia and freeze it into the ice where the USS *Jeanette* had sunk, and maybe three years later he'd pop out in Greenland. But crucially, on the way, he might pass over the top of the North Pole. No trekking, no sailing . . . just let the ice and the wind do the work for you. The only problem would be the wait. Nansen's reward for this idea was to be both hailed as a genius and derided as a madman. But he was going anyway. He raised the money and employed one of the best naval architects of the age, because the ship itself would have to be like no other ship ever floated on the ocean. And so the *Fram* was made.

The difficulty was that as water freezes, the water molecules must take their places in their rigid lattice. If the temperature sinks low enough, they will stick. And if there isn't enough room to sit in their proper place, they'll push outward on everything nearby in order to make space. Any ship frozen into the ice suffered because growing ice took up more and more space, forcing itself outwards. No known ship could resist that pressure, and no one knew how thick the ice would get in the middle of the Arctic. The *Fram* solved the problem in a brilliantly simple way. She was made to be chubby and round, only 39 metres long and 11 metres wide. She had a smooth curvy hull, almost no keel, and engines and rudder that could be lifted right out of the water. When the ice came, the *Fram* became a floating bowl. And if you squeeze a curved shape like a bowl or a cylinder from below, it will pop upwards. If the squeeze from the ice got too much, the *Fram* would just be pushed upwards to sit on top of it – or so went the theory. She was made from wood that was over a metre thick in places, and insulated to keep the crew warm. And in June 1893, she left Norway with tremendous public support and a crew of thirteen, rolling her way around the northern coast of Russia until she reached the place where the *Jeanette* had sunk. In September, she saw ice close to 78°N, and not long after that she was surrounded. As the ice first trapped her, she creaked and groaned, but as it expanded around her she rose, shifting upwards exactly as expected. Frozen in, she was on her way.

For the next three years, the *Fram* floated with the sea ice, drifting northward at an agonizingly slow 1 mile a day. Sometimes she went backwards or round in circles. The fickle freezing ice squeezed and released her, and she rose and fell in response. Nansen kept his crew occupied with scientific measurements, but got increasingly frustrated with the slow progress. When the *Fram* reached 84°N, it was apparent that she was not going to get to the pole, 410 nautical miles away. Nansen took a companion and left the ship, skiing over the ice in an attempt to go where his ship couldn't. He set a new record for the furthest north anyone had been, but his best was still 4° short of the pole. He carried on across the Arctic towards Norway, meeting a fellow explorer on Franz Josef Land in 1896. The *Fram* and her remaining eleven crew stayed the course, carried by the ice to 85.5°N, only a few miles south of Nansen's new record. On 13 June 1896, she popped out of the ice just north of Spitsbergen, exactly as originally planned.

Even though the *Fram* never reached the pole, the scientific measurements taken during her journey were invaluable. Now we knew for sure that the Arctic was an ocean and not a land, that the North Pole was hidden beneath ever-shifting sea ice, and that there really was a current that crossed the Arctic between Russia and Greenland. The *Fram* went on to carry men on two other great trips. The first was a four-year mapping expedition to the Canadian Arctic. And then in 1910 she carried Amundsen and his men to Antarctica, where they would beat Captain Scott to the South Pole. Today, she sits in her own museum in Oslo, lauded as the greatest symbol of Norwegian polar exploration. Instead of fighting the inexorable expansion of the ice, she had used it to ride across the top of the world.

The expansion of ice as it freezes is so familiar to us that we don't really notice it. Put an ice cube in your drink and it floats – that's just the way things work. But there's an easy way to see that the frozen water really is the same stuff, just taking up more space. If you put some water in a transparent glass and add some largish lumps of ice, the ice floats so that most of it is below the surface but about 10 per cent sticks up above the liquid level. You can mark the liquid level on the outside of the glass with a marker pen. The question is this: as the ice melts, will the water level go up or down? Once it's melted, all those water molecules that are now sticking up above the water level will have to join the rest of the drink. Does this mean that the water level will rise? This is proper cocktail party physics, if you're patient enough (or bored enough) at a party to spend time watching ice melt.

The answer is straightforward, and you should test it for yourself if you don't believe me. The water level will stay in exactly the same place. Once the molecules in the ice become liquid again, they can fit together more closely. This means that they'll fit perfectly in the hole that the submerged part of the ice was taking up. That bit of the ice cube that's sticking up above the water line is exactly the size of the extra volume that the ice cube has because it expanded as it froze. You can't see the atoms themselves in their lattice, but you can directly see the extra space they need when frozen.[1]

Water transforms from liquid to solid in a particular way – the

[1] The reason the space taken up by the submerged part of the ice cube is exactly the same as the space needed to accommodate the melted liquid is to do with the way that buoyancy works. The rest of the water has to support the weight of whatever is in that hole. It doesn't matter to the rest of the glass what's actually in the hole, as long as it takes up that much space. Once the ice cube has filled that space, it's got extra volume left over, and this is what sticks up above the surface.

atoms in the solid each have a fixed location in a lattice. This is called a crystal even when it isn't the gleaming centrepiece of a tiara. A crystalline material is just one that has a fixed repeating structure when it's solid, like salt or sugar. But there is another sort of solid, one without this strict positioning. These solids have a structure more like that of a liquid frozen on its way somewhere else. Even though the atomic positioning is all happening on a minuscule scale, and is far too small for us to see, we can still sometimes see the effect that it has on the object that we pick up. The most obvious example of this is glass.

I remember seeing glassblowers for the first time on a family trip to the Isle of Wight when I was about eight. I was spellbound by the smooth globules of molten glass, glowing and ballooning, constantly shifting from one beautifully bulbous shape to another. I had to be dragged away because I would happily have spent all day gazing at this wizardry, the magic of blobs flowing until they were vases. It was many years before I got around to what I'd really wanted to do: having a go at it myself. But one chilly morning this year, my cousin and I arrived at a small stone barn where they would apparently pull back the curtain and show us how the magic was done.

It started with the pool of molten glass that sat in a small furnace, glowing bright orange because it was at the horrific temperature of 1,080°C. Protected with Kevlar gloves, we obediently poked long iron rods into the pool and twisted them, so that glass with the consistency of honey wound on to the iron as we twirled it. That was the easy bit. The hard bit was everything else. Glassblowing is about controlled coaxing, and there were three main forms of persuasion that we could apply. Heating the glass up makes it softer. Holding it still lets gravity conveniently pull it downwards without you having to touch it. And if the iron rod is hollow, you can blow bubbles in the molten blob.

We took turns at practising all three, and the astonishing thing about glass is how quickly its nature changes. As the molten blob comes out of the furnace, you have to keep spinning the iron because it really is liquid; stop twirling and it will just drip onto the floor. A couple of minutes after that, we could roll the blob along a metal workbench and it felt as though it had the consistency of plasticine. Only three minutes later, you could tap it on the bench and hear it go 'ting', just as you'd expect a solid glass object to behave. The fun of glass is that you're manipulating a liquid, playing with the smoothness and curviness that liquids offer. A solid, cold bit of glass is just a liquid that was interrupted, frozen in time like a fairytale character.

Glass gets its character from the way its atoms move around each other. The most common form of glass (and what we were practising with) is soda-lime glass. It's mostly silica (silicon dioxide, SiO_2, which makes up the majority of sand), but it's also got sprinkles of sodium, calcium and aluminium in it. What makes a glass distinctive is that instead of the atoms having specific places in a regular lattice, they're all jumbled up. Each atom will be linked to the ones around it, and there won't be too much free space, but it's all quite disorderly. As the glass is heated up, the atoms jiggle about more, moving apart ever so slightly, and since they weren't in strictly regimented positions to begin with, it's quite easy for them to slip past each other. The molten glass that we took from the furnace was made of atoms with loads of heat energy, and they would easily slither over each other as gravity pulled them downward. But as it cooled in the air, the atoms would move a bit less, settling slightly closer to each other, and the liquid became more viscous.

The clever bit about glass is that as it cools down, there isn't enough time for the atoms to move into an eggbox-like regular pattern. So they don't. Glass becomes solid when the atoms are just too sluggish to move over each other any more. It's quite hard to say exactly where the line between liquid and solid really is.

The first task was to make a bauble each, which turned out to be a posh description for blowing a bubble of glass and then watching the teacher attach a loop of molten glass to the top. Blowing the bubble was hard work; my cheeks hurt afterwards as though I'd just inflated a particularly stubborn balloon. The most delicate part of the process is right at the end, when the final piece of glass needs to be separated from the iron rod. You pull and shape the glass so that there's a thin neck where you want it to break. Then you file that neck to introduce tiny cracks. And then you take it over to what was entertainingly called the 'knock-off bench', tap on the iron ever so gently, and the glass bubble breaks off. It all worked perfectly – until we were on the last one, when the newly introduced cracks weren't going to wait. The final bauble dropped off the end of the rod just as it was being finished, hit the concrete floor, and bounced. Twice. The teacher quickly scooped it up, and it was fine. But this delicate membrane of glass had *bounced*. And apparently if it had fallen just a minute or so later when it was just a little bit cooler, it would have shattered.

This is the lesson of glass. The way its atoms behave depends on its temperature. When it's hot, the atoms can flow freely over each other. Cool it just enough not to be sticky, and the atoms can press together and rebound so that the glass can bounce. A little bit cooler, and the atoms really are frozen in place. Any atom that's pushed slightly out of place opens up a crack in a fragile brittle solid, and the glass can be smashed into sharp smithereens.

Glass is intensely satisfying because it captures the curvy beauty of a liquid without you having to worry about where the liquid is going. It has the atomic structure of a liquid – a fairly disorganized mob – but it's definitely a solid. The bouncing is a giveaway: elasticity is something that solids have and liquids don't. And you can see the consequences of that structure in how the material behaves as the temperature changes.

This might be the time for a bit of myth-busting about old glass windows. It has sometimes been said that the reason that

300-year-old windows are thicker at the bottom than at the top is because the glass has flowed downward over time. This isn't true; window glass isn't a liquid and it isn't flowing anywhere. It's because these window panes were made using an incredibly ingenious method. A molten glass blob was stuck on an iron rod, and the rod was spun very quickly until the glass flowed outwards into a flat disc.[1] This disc was cooled, and cut up to make window panes. The downside of this method is that the disc will always be thicker closer to its centre. So the diamond-shaped window pane pieces were cut with the thicker bit at one end, and when it was put into the window, the thicker end was often placed at the bottom to help rain run off. The glass didn't move itself downwards, it was put there.

Our glass blobs were not allowed to cool down straight away. They were put in an oven overnight, one that would bring the temperature down slowly over the entire night until it matched room temperature in the morning. The reason for that is that even once the glass is solid, the atoms aren't absolutely stuck in place. If you heat something up, the atomic arrangement changes slightly even if the temperature change isn't enough to turn the solid into a liquid. The same thing happened as the glass blobs cooled down: the atoms shifted slightly. The reason for the oven is to allow this slight rearrangement to happen slowly and evenly over the whole structure. If it happened unevenly, unbalanced internal forces could shatter the glass. Once again, those extra internal stresses are the result of a very simple principle: the positions of the atoms may be fixed, but the distance between

[1] This is crown glass, in case you'd ever wondered where that phrase comes from. The blobby bit in the middle of very old pub windows is where the iron rod was attached. This was the cheapest section of glass, because the thickness was so uneven. Of course, these days that sort of 'character' is considered a valuable trait. As my northern family would say, 'You'd pay extra for that in a posh restaurant.' Or in this case, a posh pub.

neighbouring atoms isn't. If you heat something up, it almost always expands.

*

The world of digital measuring devices has many advantages, but it has one definite downside: we're disconnected from what the measurement really means. One of the saddest losses is the glass thermometer, an essential tool in science labs and homes for two and a half centuries. You can still buy them and I still use them in my lab, but in lots of places they've been superseded by digital alternatives. The shiny mercury thread that I remember from my childhood has been replaced by coloured alcohol, but the modern version is essentially the same as the device that Fahrenheit invented in 1709. There's a narrow glass rod with a skinny tube running all the way through the middle of it. At the lower end, the tube widens out into a bulb, a reservoir of liquid. Put this end of the thermometer in anything – bath water, your armpit, the sea – and what happens is both elegant and simple. The temperature of something is directly related to the amount of thermal energy that it has. In liquids and solids, thermal energy is expressed as the jiggling about of atoms and molecules. If you put your thermometer in the bath, you're surrounding the cold glass with hot water. The molecules in the water are moving more quickly, so they'll jostle the atoms in the glass, giving them the energy to jiggle faster as well. This is heat travelling by conduction. So when you put the thermometer in the bath water, thermal energy flows into the glass. The atoms in the glass don't go anywhere – they just fidget on the spot, vibrating from side to side. The temperature of the glass is a measure of this fidgeting, and now the glass is hotter than it was. Then the atoms in the glass bump into the liquid alcohol until it starts jiggling faster too. That's the first part – the bulb of the thermometer is heated up until it's at the same temperature as its surroundings.

When atoms in a solid vibrate faster because of extra heat, they push the nearby atoms away, just a tiny bit. The glass takes up more space when it's hot, just because its fidgeting atoms require it. This is why things expand when they get hot. But the alcohol molecules space out much more as they speed up; so alcohol expands about thirty times as much as glass for the same temperature change. Now the alcohol in the thermometer bulb is taking up more space than it was, but the only extra space is up the tube. So as the molecules in the alcohol vibrate and push each other apart, the liquid creeps up the tube. The distance it travels is directly related to the thermal energy of its molecules, and so the marks on the thermometer correspond to the amount of thermal energy in the liquid. It's beautifully simple. When the liquid in the bulb cools, the alcohol takes up less space as its molecules slow down. When the liquid heats up, it takes up more space as its molecules vibrate more energetically. So a reading from a glass thermometer is a direct measurement of atoms jostling each other.

Different materials expand by different amounts when they're heated. That's why running hot water over a stuck jam jar lid can be helpful: both the glass and the metal lid will expand, but the metal expands far more than the glass. After it's expanded, it's easier to remove; even though the difference in its size is far too tiny for you to see, you can feel the result.

Generally, solids expand less than liquids as they're heated up. The expansion is only a tiny fraction of the overall volume, but it's enough to make a difference. Next time you cross a road bridge on foot, keep an eye out for a metal strip at either end of the bridge, running across the road. It's likely to be made out of two interlocking comb-shaped plates. This is an expansion joint, and once you know what to look for, they're pretty common. The idea is that as the temperature rises and falls, the combs allow the bridge materials to expand and contract without buckling or cracking. If the bridge sections expand, the fingers of the comb are pushed further into each other; if the bridge contracts, the fingers pull back but without creating a serious gap in the road.

Thermal expansion may be elegant and useful in a thermometer, but it can have serious consequences on larger scales. One of the problems caused by our emission of greenhouse gases is that sea level is rising steadily. The current global average sea-level rise is about 3 mm per year, and it's rising more quickly as time goes on. As glaciers and ice sheets melt, water that was locked up on land is flowing back into the sea, so there's more water in the global ocean. But that accounts for only approximately half of the current rise. The other half comes from thermal expansion. As the oceans warm, they take up more space. The current best estimate is that 90 per cent of all the extra heat energy that the Earth has because of global warming has ended up in the oceans, and the extra sea-level rise is the consequence.

*

August on the East Antarctic Plateau is still and silent. While the northern hemisphere basks in summer, Antarctica spins in the darkness on the bottom of the world. On the high mountain range that stretches right across the plateau, it's almost the end of a night that has lasted four months. Very little snow falls here, but the surface ice is still 600 metres thick. The weather is calm. Heat energy is constantly leaching away into the starry night, and there is no sunlight to replace it. This deficit means that along the high mountain range, the winter temperature is regularly −80°C. On 10 August 2010, one mountainside sank to −93.2°C, the coldest temperature ever recorded on Earth.

In the ice crystals making up the snow, heat energy is stored as movement energy as the atoms jiggle about their designated location in the solid ice. So the answer to the question 'How cold can it get?' sounds straightforward: the coldest possible temperature must be the point when the atoms stop moving completely. But even at the coldest place on our planet, where there is no life and no light, there is still movement. The whole plateau is made up of atoms that are quivering, and they've got about half of the movement energy that they would have just before the ice melts at 0°C. If you could take away every last bit of that energy, they would be as cold as it's possible to get. We have a name for this temperature: absolute zero, defined as −273.15°C. It's the same for every type of atom and every situation, and it means that there is no heat energy at all. Compared to that, even Antarctica in winter, the coldest place on our planet, looks pretty warm. Fortunately, perhaps, it's very difficult to slow atoms down to a complete stop. It takes a lot of ingenuity to make sure that nothing nearby is going to give away any of its energy to your sample and spoil it all. But there are scientists who are devoting their lives to inventing extremely clever methods of removing heat energy from matter. This is the field of cryogenics, and it's opening the door to devices that are useful even in the nice warm world where we live, especially improved magnets and medical imaging technology. Most of us, however, find it very uncomfortable even to think about being really cold. So watching ducks waddle about barefoot on ice can be very puzzling.

Winchester is a pretty little place in the south of England, with an ancient cathedral and a colony of very English teashops serving hefty scones on dainty plates. It can be spectacular in summer, with colourful flowers and a bright blue sky making the whole place look picture-postcard perfect. But one year I took a friend there on a snowy winter day, and it looked even better. Bundled up in scarves and thick coats, we stomped all the way down to the end of the high street, until we reached the modest river and the soft blanket of undisturbed snow on its banks. My favourite thing about Winchester has nothing to do with stone buildings or King Arthur or scones. What I had dragged my friend all the way through town on a freezing cold day to see was far more prosaic: ducks. We crunched our way through the snow for a short distance along the river path, and there they were.

Just as we arrived, one duck on the bank waddled across the last bit of ice and hopped into the water. And then it did exactly what all the others around it were doing: it faced the stream, started paddling like mad, and reached down to dabble in the water in front of it in search of food. The river is very shallow at this point, but the water flows very quickly. There are plants growing on the bottom, just within reach, but the ducks have to paddle furiously to stay in one place in order to forage. The river in Winchester is a treadmill for ducks, and I find it endlessly entertaining. They paddle continuously, all facing the same way, and they never seem to stop.

A small child next to us looked down at her snow-covered boots, then pointed to a duck that stood on the ice on the bank and asked her mother an extremely good question: 'Why don't his feet get cold?' Her mother didn't answer because at that moment, the real comedy show began. One of the paddling ducks had got a bit too close to one of the others, provoking a burst of quacking and flapping and pecking. The funny bit was that as soon as the scrap started, both of them forgot to paddle, and so they both whooshed off downstream with the flow, quacking as they went. After a few seconds, they realized how fast they were moving, forgot about each other, and started trying to paddle back upstream to where they had started. It took a while.

The water was close to freezing, and yet the ducks probably weren't feeling the cold. Hidden beneath the water surface, a duck has an extremely ingenious way of preventing heat loss through its feet. The problem is one of heat transfer. If you put something hot next to something cold, the faster, more energetic molecules in the hot object will bump into the molecules in the cold object, transferring heat energy from the hot object to the colder one. This is why heat flow always has to go that way – molecules vibrating slowly can't give energy to those vibrating faster, but it's easy in the other direction. So heat energy is generally shared out until everything is the same temperature and equilibrium is reached. The real problem for the ducks is the blood flow in their feet. It comes from their heart, in the nice warm centre of the duck, so it's at 40°C. If that blood gets anywhere near the freezing water, there is a big difference in temperature, so it will lose its heat to the water very quickly. Then when it gets back to the body of the duck, the warm duck will give its heat away to the cold blood and the whole duck will cool down. Ducks can restrict blood flow to their feet a bit, so less blood is at risk of getting cold, but that doesn't solve the problem completely. They use a much simpler principle. It's this: the bigger the temperature difference between two objects when they touch, the faster heat will flow from one to the other. Another way of putting it is to say that the closer the temperatures of the two objects are, the more slowly heat will flow from one to the other. And that's what really helps the ducks.

As all that frantic paddling was going on, warm blood was flowing down the arteries of each duck's legs. But those arteries were right next to the veins carrying blood back from the feet. The blood in the veins was cool. So the molecules in the warm blood jostled the blood vessel walls, which then jostled the cooler blood. The warm blood going to the feet got a bit cooler, and the blood going back into the body was warmed up a bit. Slightly further down the duck's leg, the arteries and the veins are both cooler overall, but the arteries are still warmer. So heat flows across from the arteries to the veins. All the way down the duck's legs, heat that came from the duck's body is being transferred to the blood that's going back the other way, without going near the duck's feet. But the blood itself goes all the way around. By the time the duck's blood reaches its webbed feet, it's pretty much the same temperature as the water. Because its feet aren't much hotter than the water, they lose very little heat. And then as the blood travels back up towards the middle of the duck, it gets heated up by the blood coming down. This is called a counter-current heat exchanger, and it's a fantastically ingenious way of avoiding heat loss. If the duck can make sure that the heat doesn't get to its feet, it has almost eliminated the possibility of losing energy that way. So ducks can happily stand on the ice precisely *because* their feet are cold. And they don't care.

This strategy has evolved many times separately in the animal kingdom. Dolphins and turtles have a similar layout of blood vessels in their tails and flippers so that when they swim into colder water, they can maintain their internal temperature. It's also seen in Arctic foxes – their paws have to be in direct contact with ice and snow, but they can still keep their vital organs warm. It's very simple, but very effective.

Since my friend and I didn't have any way of playing the same trick, we lasted a limited amount of time out in the snow. After watching a few more high-speed duck squabbles, and expressing suitable admiration for what must be the fittest ducks anywhere in the world, we went in search of giant scones.

*

After many thousands of experiments carried out by generations of scientists, we have concluded that the fixed direction of heat flow seems to be a pretty fundamental law of physics. Heat will always flow from the hotter to the cooler object, and that's just the way it is. However, that fundamental law says nothing about how quickly the transfer has to happen. When you pour boiling water into a ceramic mug, you can carry that mug around by the handle until the water has completely cooled down, and you won't get burnt because the handle doesn't heat up very much. But if you put a metal spoon into the boiling water and keep hold of the end, you'll be squeaking with discomfort after just a few seconds. Metal conducts heat very quickly, and ceramics conduct heat very slowly. That must mean that metals are better at passing on the vibrations from the most energetic molecules. But both metals and ceramics are just made of atoms that are anchored in place and can only vibrate about a fixed location. Why would there be a difference in conductivity?

The ceramic mug shows what happens if you rely on entire atoms passing on their vibrations. As we've said, each atom nudges the one next to it, which nudges the next, and eventually the energy gets passed along the chain. This is why you can hold the mug handle without getting burnt – that method of passing the energy along is slow, and lots of it will be lost to the air before it ever reaches your hand. Ceramics, just like wood and plastics, are considered poor conductors of heat.

But the metal spoon has a shortcut. In a metal, most of the atom is locked in place, just like the ceramic. The difference is that each metal atom has a few electrons around its edge that aren't very tightly bound to it. We'll get on to electrons properly later. What matters here is that electrons are tiny negatively charged particles that sit in a swarm in the outer zone of each atom. In the ceramic they're locked in place, but in the metal they can easily be swapped between adjacent atoms. So while the metal atoms themselves must sit in their lattice positions, those free electrons can wander through the whole structure. They form a cloud of electrons that are shared between all the metal atoms, and they're extremely mobile. It's these electrons that are the key to heat conduction in metals. As soon as you pour boiling water into the mug, the water molecules pass some thermal energy to the ceramic walls, and that gets slowly passed through the mug as whole atoms bump into each other. But as soon as the hot water touches the spoon, it passes its vibrations on to both the fixed metal atoms and their electron cloud. Electrons are tiny, capable of vibrating and of zooming through a structure very quickly. So while you're holding on to the spoon, the minuscule electrons are shunting themselves about inside the metal, passing on thermal vibrations far more quickly than the whole metal atoms do. It's the electron cloud that brings the thermal energy up to the top of the spoon so quickly, heating up the rest of the metal as they go. Copper is best of all the metals for doing this, by a long way; it conducts heat about five times faster than a steel spoon. That's why cooking pans are sometimes made with copper bases but iron handles. You want the copper to share the heat out through the food quickly and evenly, but you don't want the thermal energy to find its way up the handle.

Once you have proved the existence of atoms, you naturally ask what they're up to in different situations. And that leads directly to an understanding of what heat energy actually is. We often talk about heat as though it's a fluid that can flow through or into or out of the objects around us. But really, it's just movement energy, shared around as different objects come into contact with one another. Temperature is a direct measurement of that movement energy. We can control how the energy is shared by using materials that are good conductors of heat, like metals, or poor conductors of heat, like ceramics. When you look at the control of heat and cold in our society, one system stands out above all the others for the difference it makes to our lives. We humans spend a lot of time making sure that we keep warm. But when it comes to food and pharmaceutical drugs, we have a huge invisible infrastructure for keeping things cold. Let's finish this chapter by having a look at fridges and freezers.

If a piece of cheese warms up, and its molecules speed up their dance, there is more energy in the system, and that means that there's more energy available for chemical reactions. In the case of cheese, this means that any microbes sitting on the surface can rev up their internal factories and start the process of decay. This is why refrigeration is useful. If we cool the food down, the molecules slow down, and the energy needed for the microbes to get stuck in isn't available. So cheese will last far longer in the fridge than at room temperature. Through a clever mechanism in the back, the fridge cools the air inside itself by generating more hot air outside.[1] Cold lets us preserve food, because it limits how much the molecules can change.

Just imagine what life would be like without refrigeration. It isn't just that there would be no ice cream or cold beer. You'd have to go shopping far more often, because any vegetables you bought wouldn't last. You'd have to live very close to a farm if you wanted milk or cheese or meat, and very close to the ocean or a river if you wanted fish. You'd only get fresh salad leaves when they were in season. We can preserve some foods by pickling, drying, salting or canning, but that won't help you if you want a fresh tomato in December.

Behind our supermarkets sits a vast chain of refrigerated warehouses, ships, trains and aircraft. Blueberries grown in Rhode Island can be sold in California a week after they were picked, because from the moment they were plucked from the bush until the moment they were put on the supermarket shelf, they were never allowed to take enough energy from their surroundings to warm up. We can trust that our food is safe to eat, because it's deprived of heat energy on its way to us. And it's not just food. Many pharmaceutical drugs rely on being kept cool, too. Vaccines are especially vulnerable to damage if they're allowed to heat up, and one of the issues with getting vaccines

[1] This works by using the gas laws introduced in chapter 1, the ones that control how letting gases expand and contract affects their temperature. A fridge has a motor that pumps a fluid called a refrigerant around a loop that passes from outside to inside the fridge and back out again. First, the fluid expands and therefore cools. This cool fluid is passed through the back of the fridge to the interior, where heat energy moves from the air to the refrigerant, cooling the air. Then the fluid arrives outside the fridge again, where it's compressed using a motor, and so heats up. The extra heat is lost to the air, the fluid is allowed to expand, and the cycle starts all over again.

into the developing world is that they must be kept cool all the way. The freezers and fridges that we see in our kitchens and doctors' surgeries are the last step in an unbroken chain of cold that stretches around our planet, connecting farms and cities, factories and consumers. When you or I heat up milk to make hot chocolate, that's the first time that milk has been warm since it was pasteurized just after it left the cow. And so when we trust that it's safe to drink, we are trusting the enormous chain of cold that brought it to us. The atoms in the milk have been deprived of thermal energy all the way along that chain, so the chemical reactions that would have spoiled the milk have been almost completely switched off. Preventing atoms having too much thermal energy is what keeps our food safe.

Next time you drop an ice cube into a drink, watch it melt and imagine tiny atomic vibrations sharing out energy, as heat flows from the water into the ice cube. Even though you can't see the atoms themselves, you can see the consequences of what they're up to all around you.

7

Spoons, spirals and Sputnik

The rules of spin

ONE OF THE NICE things about bubbles is that you know where to look for them: at the top. They're either on their way there, wobbling upward through fish tanks or swimming pools, or nestled in with the crowd on top of champagne or beer. Bubbles reliably find their way to the highest point of the liquid they're in. But next time you stir a mug of tea or coffee, have a look at what's going on at the surface. The first odd thing that happens is that as you move the spoon around in circles, the surface of the tea develops a hole. As the liquid whirls around, the middle of the tea sinks and the edges rise up. And the second odd thing is that the bubbles in the tea are spinning quietly at the bottom of the hole. They're not at the highest point of the tea, at the edges. They're hiding at the lowest point on the surface, and they stay there. If you push them away, they find their way back. If you make new bubbles at the edges, they spiral into the centre. Odd.

When I start to stir my tea, I'm pushing on the liquid with the spoon. I push it forwards, but there's only so far that it can go before it meets the side of the mug. If I did the same thing with a spoon in a swimming pool, the water in front of the spoon would move forward, and it would keep going forward until it mixed in with the rest of the pool. But in the mug, there's no room for that to happen. Even though the side of the mug isn't going any-where, it can still push back on any liquid that bumps up against it. It's a wall, and tea can't pass through it. Since the tea can't go in a straight line, it starts to move around the cup in a circle. But as that's going on it's piling up against the walls because only the side of the mug can push back. The tea will keep trying to go in a straight line, and it only moves around in a circle because it's being forced to curve.

This is the first lesson about spinning things. If you suddenly freed them of their restriction, they would just keep moving in the same direction they were going in at the moment of release. Imagine a discus thrower, spinning around while holding on to the discus. After a few rotations, the discus is going incredibly quickly, but it stays on its circle because it's being tightly held. The athlete must continually pull it back towards the centre of the spin, and that pull is along the line of his or her arm. The second that they let go, the discus travels forward in a straight line, with exactly the direction and speed it had before the release.

As I'm stirring my tea, the hole develops because each bit of tea tries to move in a straight line, but that makes it push up against the cup sides and so there's less left in the middle. When I stop stirring, the hole remains because the liquid is still spinning. As the whirling slows down, it takes less of a push to keep the tea going in a circle, and so there's less of a pile-up at the sides. You can see all this in a liquid because it's free to move, so it can alter its shape.

And at the centre of the circles, the bubbles are spinning away. What their presence in the middle tells you is that it's the least favourable place to be. When a glass of beer is sitting on a table, the bubbles move to the top because the beer is winning the competition to get closest to the bottom. And it's the same for the mug of tea. The bubbles are in the middle because the tea is winning the competition to move out to the sides. The liquid is more dense than the gas, so the gas drifts into the space left over.

Our civilization is full of things that spin – clothes dryers, discus throwers, flipped pancakes and gyroscopes. The Earth itself spins as it circles the Sun. Spinning is important because it lets you do lots of interesting things, sometimes involving enormous forces and oodles of energy, all without actually going anywhere. The worst that can happen is that you end up back where you started. The bubbles in the tea are just the start. The same principle also explains why you wouldn't launch a rocket from Antarctica and how doctors measure whether you have enough red blood cells. Spinning could also play an important role in our energy grid in the future. All of those possibilities come from a restriction: the one thing you can't do when you're spinning is to travel in a straight line.

If you're zooming around in a circle, there must be something either pulling or pushing you inwards, forcing you to change direction constantly. That applies to anything that's spinning, whatever the situation. If that extra force is taken away, you just continue in a straight line. So if you want to travel in a circle, you must have something to provide an extra, inward push. The faster you're going, the stronger that push has to be, because the faster you need to curve, the more force it takes. Spectator sports love a good racetrack; they have the same benefit as anything else that spins. You can achieve huge speeds without really going anywhere, and certainly not anywhere the paying audience can't see you. To make sure that they get enough of an inward push to stay on the track, some sports have taken racetrack building to fairly extreme lengths. Indoor track cycling is the prime example of this. But it wasn't the lengths that terrified me when I tried it . . . it was the steepness.

I've been an enthusiastic cyclist all my life, but this was a very different kettle of fish. The inside of London's Olympic velodrome is shiny and vast and oddly quiet. You pop up into the middle of the stillness, and they issue you with a lean, mean-looking bike with a single gear, no brakes and the most uncomfortable saddle I've ever had to sit on. When the group for the beginners' session was assembled, we clopped outwards to the track and held on to the rail while we clipped into the pedals. The track seemed enormous. There are two longer straight sides, and then the sweeping banked sections at each end that towered over us. They are so steep (43° in places) that it seems as though the designer had really wanted to build a wall. Cycling looked like entirely the wrong thing to be doing here. But it was too late for our little group now. The track was waiting for us.

First of all, we were sent around the flat oval that sits inside the main track. The surface was beautifully smooth, and the bikes made a lot more sense. Then we were instructed to venture outwards on to the light blue strip with the first slight gradient. And then, feeling slightly like baby birds being pushed out of their nest to learn to fly, we were told to face the main track.

There was a nasty surprise straight away. I had thought the banking would be gradual, but it isn't. The gradient at the bottom is pretty similar to the gradient at the top. As soon as you stray outwards on to the race surface, you're cycling across a pretty significant slope. Pedalling faster seemed like a good idea, but that was only because I was forcing my brain to let logic make the decisions while it busily pretended that instinct didn't exist. I forgot how stupidly uncomfortable the saddle was after the first three laps. Round and round we went, like demented hamsters on a gigantic wheel, pausing occasionally so that the instructors could check on us. Twenty-five minutes in, I was still terrified, but I was learning.

The game here is that you want the bike to lean inward so much that it's perpendicular to the track. The only way to do that without slipping down the slope is to be going very quickly, because then you're like the swirling tea. The bike wants to continue in a straight horizontal line, but it can't because the curved track is in the way. That push-back from the track is providing the inward force that keeps you going in a circle. The bike is pushing into the track so hard that when you add that push to gravity, it's as if gravity has changed direction. Now you're being pulled into the track, instead of downward to the centre of the Earth. The faster you cycle, the more you change the direction of the effective gravity. It still feels as though you're cycling on a wall, but at least it's one that you're glued to by something that feels familiar.

I understood the theory, but the practice was a bit different. For a start, there's no resting. You can't stop pedalling because you can't freewheel. If the wheels are going round, your legs are going round and that's just the way it is. On a few occasions, I instinctively paused as I would if I needed a couple of seconds' rest on the road, and I was rewarded by a gigantic kick of adrenaline as the bike bucked me up out of the saddle. You can't freewheel on these bikes at all. You have to keep going, however much your legs are burning. If you slow down, you slide down the bank. I felt some new respect for the athletes who do this on a regular basis. And then there are all the other people on the track to contend with. If you move up to overtake someone, you are taking a longer path, so you have to increase your speed a lot even to have a chance. I was quite happy not to do too much overtaking.

The lesson from all this was that if you're doing things right, the steeper slopes will give you a stronger inward push. And the reason that you need that push on the ends but not along the sides is that the semicircular ends are where you're changing direction. The more quickly you change direction, the bigger the push you need to make it happen. If you tried to cycle this quickly on a flat track of the same shape, you'd skid out to the sides – tyre friction alone can't supply that inward push. The velodrome is what happened when the cycling world refused to allow its need for indoor speed to be limited by friction.

If you've ever wanted to know what it feels like to be a penny rolling down a whirlpool-shaped charity donation funnel, this is the way to do it. At the end of an hour I was properly fired up on adrenaline, and really glad that it was time to stop.[1] The scary thing about the effective gravity pulling me into the track was the knowledge that if I slowed down suddenly, it would change back. And gravity pointing downwards is a pretty uncomfortable thought when you're cycling on a 43-degree wall.

The cyclist is being pushed inwards by the track in the same way that the ground is pushing up on us all the time. If the ground underneath you suddenly vanished, you would fall because gravity is pulling you downwards. So the ground itself is pushing back on us, to counter gravity's downward tug. Cyclists feel the track both pushing them upwards and pushing them inwards. Overall, that will feel as though gravity is pulling them down and outwards.

There's a track cycling event appropriately called the 'flying 200-metre time trial'. I reckon it must feel like flying, even though it's called that because they're already up to speed before the clock starts. The world record at the time of writing was set by François Pervis and it's 9.347 seconds. That's 21 metres every second, or nearly 48 mph. For him to spin around the end of the track at that speed, the track has to be pushing him inwards almost as hard as the floor is pushing him upwards. François was glued to the track by a force almost twice as strong as normal gravity.

As we saw in chapter 2, a constant background force like gravity is useful for all sorts of things, although some of them (like separating out cream) take ages. But spinning offers us an alternative. You don't need to move to a new planet to reap the benefits of increased gravity. Cyclists can almost double their effective gravity at the top of a track, but even the best track

[1] I should emphasize that I did enjoy it, mostly, and I would definitely recommend the experience. As long as you're a confident cyclist to start with, it's a great way to appreciate rotational physics in a very visceral way.

cyclists in the world can 'only' get up to about 50 mph. In theory, you could just keep spinning faster and the forces on you would keep getting stronger.

Remember gravity helping the droplets of cream to separate from the rest of the milk and rise to the top of the bottle in chapter 2? If the force pulling the milk downward is only as strong as gravity, it takes a few hours for the fat droplets to separate out. But if you put the milk in a long, spinning tube and whizz it around very quickly, the outward pull is so strong that the cream droplets will separate out in just a few seconds. This is how all our cream is separated from the milk these days – they don't just let it sit and wait for it to sort itself out. Modern food production hasn't got the time for that. Spinning something around generates a pull that can be as strong as you like, as long as you can spin fast enough. This is what a centrifuge is: a spinning arm that can hold on to something, pulling it inwards to make it spin, and making the object feel as though it's being squashed against the outer side by a very strong force.

You can make those internal spinning forces so strong that things that would never separate out under gravity alone can be teased apart. For example, if you ever have a blood test for anaemia, laboratory technicians will put a sample of your blood in a centrifuge and spin it so fast that it experiences an outward force perhaps twenty thousand times as great as gravity. Red blood cells are far too small to separate out under gravity in any normal circumstances, but they can't resist the forces generated by the centrifuge. Under those conditions, it takes only five minutes for almost all the red blood cells to be pulled outwards from the centre of the centrifuge, towards the bottom of the tube. They're more dense than the liquid they're in, so they win the race to the bottom. Once they're all there, the tube can be lifted out and it's possible to directly measure the percentage of your blood that is made up of red blood cells, just by measuring the thickness of the lowest layer. This is a simple test that can indicate a range of health problems, and it's also used to check for blood doping in athletes. If it wasn't for the forces generated by spinning, this measurement would be much harder and much more expensive to carry out. And these forces can be applied to much larger things than blood samples. One of the biggest centrifuges in the world is designed to spin an entire human being.

Many people envy astronauts for their adventures: their amazing views of our home planet, all the technical toys they get to play with, the accumulation of fabulous stories to tell, and the accolades of having one of the rarest and most hard-earned job descriptions in the world. But ask most people what they envy the most and you almost always get the same answer: weightlessness. All that floating about without 'up' or 'down' being a problem sounds both highly exciting and very relaxing. So it might seem slightly strange that astronauts in training need to be just as well prepared for the opposite problem: forces that far exceed gravity. The only current way to get to space is to sit on top of a rocket that's accelerating pretty quickly. And it's even worse on the way back down: re-entry into the Earth's atmosphere can generate forces four to eight times as strong as gravity, the sort of forces that a fighter pilot doing tight turns at high speed might have to deal with. If you feel slightly queasy as a lift accelerates, this might not be for you. Depending on the direction of the additional g-forces, extra blood will be pushed towards or away from your brain, possibly even bursting the tiny capillary blood vessels in your skin. The details aren't necessarily pleasant. But humans can not only survive these forces, they can also work while subjected to them (as you have to if you're piloting a spacecraft back to Earth), and they do both better if they're used to it. So a way has been found to train them.

All current astronauts and cosmonauts will spend considerable periods of time at the Yuri Gagarin Cosmonaut Training Centre in Star City, just to the north-east of Moscow. Among the lecture halls, medical facilities and spacecraft mock-ups sits the TsF-18 centrifuge. From the centre of a huge circular room, the arm of the centrifuge stretches 18 metres outwards. The capsule at the end can be changed depending on what's needed on any given day. The tests that any budding astronaut has to pass involve sitting in the capsule while the arm rotates once every two or four seconds, which doesn't sound like much until you calculate that the capsule itself must therefore be travelling around at either 120 or 60 mph. Once they've shown that they've got the right stuff, astronauts can practise working in these conditions, and are constantly monitored to check on how their bodies are responding. And it's not just astronauts – test pilots and fighter pilots can also train here. The centre even offers the experience to members of the public who can afford to pay for it. Be warned, though: the only thing about it everyone seems to agree on is that it's very uncomfortable. But if you want to experience a consistent very high force, taking a spin is the way to do it.

The centrifuge is one way of exploiting the forces generated when something spins: by taking advantage of the ability to generate a very strong force in a single direction, and treating it like artificial gravity. But there is also a second way of employing forces from spinning. The tea and the cyclist and the astronaut were all confined – they were all forced to move in a circle because there was a solid barrier pushing back on them, preventing them from moving outwards. But what if you're spinning and there's nothing external to trap you on a fixed circular path? This is a pretty common scenario. Rugby balls, spinning tops and frisbees all spin without anything external pushing them inwards. But the best way of seeing what's going on is far more fun, and also edible: pizza.

To my mind, the perfect pizza should have a thin crispy base, the vital but understated foundation that lets the toppings shine. Raw pizza dough starts out as a rotund blob, a living lump that needs to be kneaded and nurtured to bring out the best in it. Transforming the blob into a delicate sheet without breaking it is an essential skill for a pizza-maker, and some go a step further, taking the basic skill and turning it into theatre. The chefs who toss pizzas have mastered the trick of letting the spinning do the work for them. Why push and prod each part of the dough with your fingers when you can just let physics sort out all those messy details? Especially when the flying disc gives you the mysterious aura of a dough wizard.

Tossing pizza dough has evolved into a proper spectator sport of its own; there's now a world championship every year. There are even those who call themselves 'pizza acrobats', whose party piece is keeping a constantly spinning pizza base (or two) flying and somersaulting around their body for several minutes at a time. No one seems to eat pizza made from such well-travelled dough, but it definitely looks impressive. However, there are plenty of pizza chefs out there who spin their pizza dough briefly without making a cabaret act of it, and who have every intention of turning it into someone's dinner. What is the spinning actually doing?

Some pizza-mad friends of mine recently took me to a very friendly restaurant with an open kitchen, and I asked whether I could watch someone spinning pizza dough. The young Italian chefs giggled a bit, but then gathered around the one who was brave enough to volunteer. Half embarrassed and half proud to show off, he patted a ball of dough to flatten it slightly, then picked it up and with a slight flick of the wrist, sent it twirling into the air.

What happened next happened very quickly. As the circle of dough left his hand, it was suddenly free of anything external pulling and pushing on it. It's helpful to think about a single point on the edge. It's travelling around in a circle, but only because the rest of the dough is stuck to it, pulling it inwards. That inward pull is always necessary for something to rotate. In the case of the cyclist, the track is constantly pushing the bike from the outside, so that the cyclist must curve inwards towards the centre instead of continuing on a straight line. For the pizza dough, it's a pull from the middle that makes the edge of the dough curve around towards the centre. Either way, there has to be a force directed towards the middle of the spin. But dough is soft and elastic. If you pull on it, it stretches. The middle of the dough is pulling the edge inwards, but that means there's a pulling force across the dough. And so the dough has to stretch. When any solid object spins, the spinning generates forces inside it that you can't see. The internal pull that's keeping the pizza together is also stretching the dough, and the edge is getting further and further from the centre. The brilliance of this for a pizza chef is that the internal pull is smooth and symmetrical. All of the pizza is spinning, so all of it stretches out away from the centre.

You can feel these internal pulling forces yourself sometimes. If you hold a bag containing a reasonably heavy object out horizontally, and spin yourself around, you'll feel a pull trying to stretch your arm. This is the inward pull that's keeping the bag spinning in a circle. Fortunately for you, your arm is much less stretchy than pizza dough, so it stays the same length. But the longer your arm and the faster you spin, the more of a pull you'll feel.

So as the pizza dough was spinning in the air, the same pull that was keeping the edge moving around in a circle was gradually stretching the dough outwards. I reckon the dough was in the air for less than a second, but it was a very thick pancake when it went up, and a beautiful thin smooth circle when it came down. The chef kept it spinning and sent it up again, but this time the internal pulling forces were so strong that the dough tore itself apart in the middle, and what came down was sad and raggedy. The chef grinned sheepishly. 'That's why we don't normally do it,' he said. 'The dough that makes the best pizza is too soft to spin, so we have to stretch it by hand on the board.'[1] It turns out that the dough used in the acrobatic competitions is made using a special recipe so that it's stretchy and strong, but doesn't necessarily produce the cooked pizza with the best texture. Right at the edge of a pizza, the internal pulling force could be five to ten times as strong as gravity, which is why the pizza base stretches much more quickly when you spin it than it does if you just hold it up and let it droop under its own weight.

A spinning pizza base is lovely to watch because it changes shape in response to forces that are entirely hidden away inside itself. Spinning anything generates a pulling force from the centre to the edge – the same is true in a spinning rugby ball or frisbee – but you'd never know about it in these solid objects because they're strong enough to resist being stretched. Or at least, they're stretching so little that you can't tell. But everything will stretch a tiny bit. Even the Earth itself.

*

Our planet is constantly spinning as it travels around the Sun. And, just like the pizza dough, it's stretched by the forces that are pulling each bit of it back inwards, keeping each bit of rock travelling in a circle. Fortunately for all of us, gravity is strong enough to prevent any consequences as extreme as they are for the dough, and the Earth stays fairly spherical. But it does have what is helpfully called an 'equatorial bulge', which sounds like a euphemism for having eaten too much cake. If you stand at the equator, you're 21 km further away from the centre of the planet than someone at the North Pole is. Our planet is held together by

---

[1] I'm sure that the pizza-lovers out there will all have their own strong opinions on what makes the best pizza dough and how to shape it. I can vouch from personal experience that the pizza produced by this restaurant was superb. But don't write me letters if you disagree with the chef's conclusions!

gravity but shaped by its spin. And so even though Mount Everest is the tallest mountain on Earth, the top of Everest is not the furthest point from the centre of the Earth. That accolade goes to Chimborazo, a volcano in Ecuador. Its summit is only 6,268 metres above sea level (Everest is 8,848 metres tall by the same measure), but it's sitting right on top of the equatorial bulge. So, when you're standing on top of Chimborazo, you're a little over 2 km further from the centre of the Earth than anyone who has just struggled to the top of Everest. Pointing that out when you both get home probably won't make you popular, though.

Overall, the forces generated by spinning can be useful in two ways. The pizza is one – spinning something without confining it generates a pull inside the object as it tries to hold itself together while it spins. The cyclist is the other – if you do put a wall in the way, confining whatever is spinning with something that pushes back, you can generate a strong consistent gravity-like force on that object. But the common theme is that an inward pull or push has to come from somewhere. If that inward force ever goes away, the object can't stay on its circular path.

Only a solid object can hold itself together like the pizza dough. Liquids and gases aren't stuck together like that.[1] This distinction is fantastically useful if you have both solid objects and liquids mixed together, because now you can separate them out. The brilliance of a spin dryer for clothes is that the clothes are trapped inside the drum, and the drum is pushing them inwards so they have to keep going round in circles. But the water tucked away in the clothes isn't held in position. Since it's free to move, it can keep moving outwards through gaps in the material. It will only travel in a circle if it gets an inward push from something solid. Otherwise, it will gradually wriggle its

---

[1] Unless it's a drop of liquid so small that surface tension can do the job. But a droplet has to be very tiny for that to be enough.

way away from the centre, and when it meets a hole in the drum, it'll go flying out sideways, free of the circle completely.

When you spin something around and then let go, you start by pulling on it with exactly the right inward force to keep it going round in a circle, and then you suddenly take that force away. When there's no inward pull, there's no reason for the object to keep going round in a circle. So it just sails off in a straight line. This principle revolutionized medieval warfare in Europe and the eastern Mediterranean, enabling engineers to build giant siege engines that could batter down stone fortresses. And I have used it to launch wellies, but not quite as effectively.

At the end of my PhD viva, just after they'd told me that I'd passed, the external examiner smiled across the table and asked what I was going to do with the rest of the afternoon. He was clearly expecting parties and pubs and inebriation to be high on the agenda. He was not expecting me to say that I was about to go cycling out into the Cambridgeshire countryside to see whether I could find a farmer who might lend me an old tractor tyre or two. I explained that I was making a contraption to throw wellies with, that I had to make it out of scrap material, and that I had to finish it within the next week. The examiner's brow wrinkled, his eyebrows waggled uncertainly, and then he smoothly pretended he hadn't heard and asked me what plans I had for a job. But it was true. I had agreed to be part of a rare all-female team taking part in a *Scrapheap Challenge* roadshow, and the challenge was to build something that could compete at throwing wellington boots, to be put through its paces at the Dorset Steam Fair. There were three of us, we had no money and very little time, and as far as I could see the only option available was to use an ancient and very effective technology: the trebuchet.

A trebuchet is an extremely ingenious device that was developed over many centuries with the input of several civilizations: the early Chinese, the Byzantine and Islamic empires, and finally western Europe. When it came of age in the eleventh and twelfth centuries, it proved itself a monstrous lumbering brute, capable of demolishing castles that had previously been thought impregnable. A trebuchet could hurl 100-kg rocks over hundreds of metres. Siege engines like this contributed to the disappearance of motte-and-bailey castles (strategically useful but made only of wood and earth). Solid stone was the only defence, and so stone fortresses became the norm.

The benefits of the trebuchet were the same for me and my team as for the medieval warmongers: it's mechanically simple and extremely effective. We borrowed scaffolding poles from a local building site, dug around in the college skip for stuff to make a sling from, persuaded the technicians in the Cavendish Laboratory to let me have a 5-metre-long metal beam, gathered all this stuff together at the top of the college playing fields and set to work. Churchill College in Cambridge had been my home for nearly eight years by then, and the college staff were used both to me and to the sudden appearance of novel contraptions. Looking back, I'm still astonished at (and extremely grateful for) the cheerful acceptance that met any of the students whenever we had a new daft idea. At the opposite end of the playing field that week, someone else was testing a stratospheric balloon to send a teddy bear into space.

The basic structure of a trebuchet is very simple. You build a frame that gives you a pivot point which is perhaps 2 or 3 metres off the ground. Then you attach one long beam to that, like a giant see-saw, but you position the pivot point so that there's much more of the beam on one side than the other. Now you've got an A-frame that looks as though it's got a long stick laid across the top. The long end is the one that starts touching the ground. You attach a sling to the long end, and lay the sling on the ground underneath the frame. The first time we assembled it all, it was a beautiful sunny day, perfect for launching anything.

Then we hit a problem. The beautiful thing about a trebuchet (unless you're the one that's about to get a rock thrown at you) is that it uses gravity to spin the see-saw and the sling. You attach a heavy weight to the short end of the see-saw, and then as you drop the weight, it pulls the see-saw on your side down very quickly. The whole beam spins around the pivot, tracing out a vertical circle, and the sling also spins around the other end of the beam. So you've got lots of very fast rotation, and the project-ile in the sling is spinning around the pivot because it's being pulled inwards by the sling. So far, so good. The first task was to get to this point, but we couldn't find a weight that was heavy enough to move everything. I offered to swing from the beam myself as a human weight, but even I wasn't quite heavy enough. We were stumped. That night, I spent a while pouring out my frustration to another set of friends, batting away their sugges-tions that I should just eat more cake. Then one of them offered me his scuba-diving weights. So the next day, I rigged myself up with a belt carrying 10 kg of diving weights, and we tried again. This time, it worked perfectly. I swung under the pivot, the see-saw swung over the top, and the sling swung over the top of that. Everything was spinning. Now it was time for the next step.

The sling is only held in place by a small loop, and the trick to it all is that when the sling is almost at its highest point, the loop slips off. The sling is effectively broken. That means that the force that was pulling the projectile inwards and keeping it on the circle has vanished. Now the situation has changed. At this moment, the projectile in the sling is travelling forwards and upwards very fast. As soon as it's free of the inward force, it just keeps going in a straight line. Since it was travelling forwards and upwards before, it keeps going forwards and upwards. But it doesn't go directly outward from the centre of the spin. It car-ries on going sideways, as if following a line that sat on top of the spin circle. That was the theory. We put a shoe in the sling and lined everything up. I faced away from the playing field and swung down on the see-saw. The other end of the see-saw swung up, dragging the sling up, around and over the pivot. At exactly the right moment (first time!) the sling released, and the shoe went flying over the top of my head, out on to the playing field. I wouldn't ever want to do that with a rock, but the shoe proved the point perfectly. Our contraption could at least throw a wellington boot, and in the time we had, that was the best we could do. After a bit more practice, we took our frame apart ready to transport it to the competition the next day.

Arriving at the Dorset Steam Fair poked a large pin in our buoyant bubble of confidence. Every other team consisted of middle-aged men who had spent months in garages building gloriously decorated contraptions for throwing wellies. Our small pile of scaffolding poles and discarded carpet, collected over just a few days, looked sparse and unloved. But we put a brave face on it and put it all back together. The competition officials (also middle-aged men) came round to look at it. 'It's silly to swing from it,' one said. 'You should do what the medieval warriors did, and just pull the lever down with a piece of rope. That'll work far better.' My protests that the counterweight was the invention that had led to the success of this device went unheard. The reason it never made it as a powerful siege weapon before the eleventh century is precisely because people tried to do it by manpower. But the officials stuck their hands in their pockets, opined that pulling on a rope was a far better idea, hinted that we enthusiastic but inexperienced females should be grateful that we were getting extra help from them, and didn't leave until my team-mates had given in and agreed with them. There was no time to argue. The time of the competition had come.

The first challenge was to throw as many wellies as possible over a line about 25 metres away in two minutes. The top five teams would then carry on to the next stage of the competition, which was to see who could throw the furthest overall. The clock started. The three of us heaved on the rope, turning the see-saw and flinging the sling. But the first welly barely cleared our heads. We couldn't pull downwards fast enough to make the see-saw rotate properly. We tried again. And again. After about a minute, I convinced my team-mates that this wasn't working, and we set ourselves up for the original idea. I put my diving weights on, jumped off the small filing cabinet we were using as a platform, swung underneath the pivot, and wheeeeee ... the first welly went soaring over my head and over the line. Next one. Welly in the sling, up on filing cabinet, swing down, whoosh! Next – but the whistle had gone. Our time was up. Two wellies over the line wasn't enough. We wouldn't get through to the next stage. The middle-aged men commiserated with us. Better luck next time. I hid from the official who had suggested the rope, because I was so cross with him. It *worked*! Our simple structure of scaffolding and carpet and elegant physics worked, and it had worked the way I'd said it would. We could have competed against the lovingly painted, intricately made garage beauties! But we had been scuppered by the last-minute change of plan.[1] Most of the other competition entries were based on far less efficient methods. They might have been colourful, but we had physical efficiency and simplicity on our side.

So my personal success with trebuchets is a bit limited, but eight hundred years ago this elegant idea revolutionized warfare. Being able to throw heavy rocks with high accuracy meant that you could thump the same bit of the castle wall again and again, until it yielded. For two centuries or so trebuchets got bigger and better, and were given names like 'God's stone thrower' and 'Warwolf'. Each one took vast amounts of timber to construct, but being able to chuck another 150-kg rock at your enemies every few minutes was worth it. Spinning the rock and the sling around an axis lets you build up to a very high speed in a very short space of time. You don't want the spinning to

---

[1] Bitter about this, ten years on? No, no ... what would possibly make you think that?

202

continue – you're just using it as a way of reaching a high speed. Once the projectile is going fast enough, you remove the inward force at the moment when its direction is perfect. And off it goes, soaring out in exactly the direction it was released in. Until gunpowder became reliable enough for the cannon to be more of a weapon than a liability, in terms of destructive efficiency a trebuchet was as good as it got.

*

Lots of things are spinning. For example, right now, you and I are spinning. We're going round the axis of the Earth once a day, although we can't feel it because the Earth is so big that we're only changing direction slowly. If we were at the equator, our sideways speed would be 1,040 mph. In London, where I'm writing this, I'm speeding sideways at 650 mph because we're closer to the spin axis. But if we all live on a massive spinning planet, and if a loose object at the surface of a spinning thing will whizz off in a straight line when you let go of it, why are we all still down here? The answer is that the inward pull of gravity is strong enough to prevent the planet letting go of us. In fact, even when you're in orbit, the planet hasn't actually let go of you. And when you're on your way up there, that extra speed you've got because of the Earth's spin can be extremely useful.

On 4 October 1957, a diminutive metal sphere named Sputnik chirped out the first sounds of the Space Age, and the world listened open-mouthed. Earth's first artificial satellite was a huge technological achievement. Sputnik orbited its home planet once every ninety-six minutes, and each time it went past anyone with a short-wave radio could hear its distinctive 'peep . . . peep . . . peep . . .' America had woken up that morning happily complacent in the knowledge that it was the greatest nation on Earth. It went to bed shocked that maybe it wasn't. Within a year, the Soviets had sent up Sputnik II, a bigger satellite carrying a dog named Laika. The panicked Americans hadn't sent anything into space, but they had launched NASA, the National Aeronautics and Space Administration. The Space Race had begun in earnest.

But what was the real achievement of Sputnik? It wasn't just about going up; anything close to something as big as a planet has to live up to the maxim 'what goes up must come down'. The trick of putting satellites into orbit starts with what goes up, but the real skill is in delaying their coming down for as long as possible. Sputnik hadn't escaped the Earth's gravity. That wasn't the point. Douglas Adams summed it up perfectly, and accurately, with the minor caveat that he was talking about flying and not orbital space flight: 'The knack lies in learning to throw yourself at the ground and miss.' Sputnik was permanently falling towards the Earth. It just kept missing it.

Sputnik was launched from the deserts of Kazakhstan, a place that is now the site of the Baikonur Cosmodrome, a vast space launch facility. The rocket that carried Sputnik powered upwards through the thickest part of the atmosphere and then turned sideways, accelerating horizontally around the curve of the Earth. By the time the last parts of the rocket fell away, Sputnik was whizzing around the planet at about 8.1 kilometres every second, or 18,000 mph. This is where the effort goes when you're getting into orbit – it's mostly about going not upwards, but sideways.

The little metal sphere hadn't escaped gravity at all. In fact, it needed gravity to be there, to make sure that it stayed in orbit and didn't just carry on and leave the Earth behind. As it zoomed along at this fantastic speed, the Earth was pulling it downwards with almost as much gravity as there is on the ground.[1] But

---

[1] Sputnik had an elliptical orbit, so its height above the surface ranged from 223 km to 950 km. That gives you a gravitational pull of between 93% and 76% of the value at the Earth's surface.

because Sputnik had such a huge sideways speed, by the time it had fallen a little way down towards the Earth, it had gone so far forward that the Earth had curved away beneath it. And as it kept falling, so the Earth's surface kept curving away. This is the beautiful balance of being in orbit. You're going sideways so quickly that you fall towards the ground and miss. And because there's almost no air resistance, you can just keep falling and missing, as you go round and round.

To get into orbit, you have to be going sideways fast enough to make this balance work. And Kazakhstan already has a pretty significant sideways speed because it's whizzing around the Earth's axis once a day. The further you are from the spin axis, the faster your sideways speed is. So by launching from somewhere close to the equator, you're giving yourself a pretty significant head start. A sideways speed of about 8 kilometres per second is needed to make low-Earth orbit work. Kazakhstan is whizzing sideways at about 400 metres per second (894 mph). So if you launch eastward, with the spin of the Earth, just starting in Kazakhstan instead of at the North Pole means that 5 per cent of the work is done for you.

In the spin dryer, the outside of the drum pushes the clothes inwards, so they can't escape. In the velodrome, it was the alarmingly sloped track that was pushing me inwards. And for Sputnik, the tiny peeping herald of humanity's first adventure into space, gravity was doing that job. Everything that spins needs something to be pulling or pushing it towards the centre of spin all the time. For the clothes in the spin dryer just as for Sputnik, if that force had vanished, they would have kept going in a straight line.

So gravity definitely still matters a few hundred kilometres above our heads. But surely the whole point of being in space is that you get to be weightless. What about all those astronauts drifting about in zero gravity, trying desperately not to spill anything because it'll float around for days? Today, the International

Space Station is orbiting above our heads. The astronauts who live on board this huge scientific facility proudly state that they are flying on particular missions, and I don't begrudge them that. It sounds a lot less exciting to say that you're going to spend six months falling. But they're not flying and they are falling. Just as Sputnik was falling towards the ground and missing, so are the astronauts and the space station.

While you're in free fall, you can't feel gravity because there's nothing pushing back. Since the astronauts can't feel anything pushing back, they can't tell that gravity is there. It's just like the moment when an elevator starts to descend, and you briefly feel lighter – the floor isn't pushing back on you as hard as it was. If the elevator were to fall as fast as it could, through a very deep elevator shaft, you'd feel weightless, too. In orbit, you haven't escaped gravity. You've just found a way to ignore it. But while you can't feel it, it's still there, its inward pull keeping you spinning around our planet.

Rotation is useful in all sorts of ways, but there are times when it's just a nuisance. For example, why is it that toast falls butter-side down? You've just whisked the hot toast out of the toaster and applied a layer of butter which is now starting to melt. All it takes is a moment of distraction while you reach for your tea, and you knock the toast towards the edge of the table. It teeters on the edge, and the next thing you know it's on the ground, face down. The lovely melting butter is now decorating your floor. It's a nuisance to clean up, made worse by the feeling that this must be the universe getting back at you for something. Why does it have to happen in the messiest possible way? Why does it turn over like that?

This is a real phenomenon. Various people have conducted experiments where they have patiently pushed toast off tables many times, and it really does fall butter-side down far more often than it falls butter-side up. It depends a bit on how the fall starts, but generally, this is the way the world works and we're stuck with it. And it's got nothing to do with the additional weight of the butter. Most of the butter soaks into the middle of the toast, and even if it didn't, it's only adding a tiny amount to the total mass of the bread.

The first question is, why does it turn over at all? It all happens so quickly that it's hard to see (and anyway, if you'd been looking at the toast, you probably wouldn't have knocked it off the table to start with). You can watch it happening if you're happy to sacrifice a bit of toast[1] or even a placemat or book that's about the same size. Put your sacrificial slice of toast flat on the table right next to the edge and nudge it towards the precipice. Just as the halfway point of the toast is over the edge of the table, two things happen. One is that the toast starts to pivot around the table edge like a see-saw. The other is that the slice starts to slide outwards without any more nudging. The toast will now take care of itself. Slip, spin, splat.

So the rotation starts when the toast is just over halfway off the table. The key to it all is that at this moment, for the first time, less of the toast is supported by the table than is hanging off the edge. Gravity is pulling down on all of the toast. The table pushes back up, but the air can't. It's all about balance, just like a see-saw. The halfway point is the time when the gravity pulling down on the overhanging side is just barely enough to lift up the toast that's still on the table. Physicists call the position of that halfway point the 'centre of mass', which just means that a see-saw pivoted at that point would be perfectly balanced.

The moment you realize the toast is falling is the moment it's too late to do anything about it. Once the toast has slipped off the

---

[1] For the sake of domestic harmony, it is probably better not to butter this particular experiment. If you insist on replicating the real situation, at least put some newspaper on the floor where it's going to land, or whatever it is that does that job in a paperless society. Protecting surfaces is one function of a newspaper that a fancy tablet computer is never going to replace.

table, it's going to take a fixed amount of time to fall. If your table is about 75 cm high, it will take the toast just less than half a second to hit the floor. But once the rotation has started there's no reason for it to stop, and the toast keeps turning as it falls.[1] Since gravity is always the same, and tables are pretty much always the same height, the toast always has the same spinning speed. In 0.4 seconds, it will turn 180 degrees. Since the butter started on the top, it ends up on the bottom. The physics is pretty much the same every time, so the outcome is pretty much the same every time. Toast falls butter-side down.

Interestingly, there is really only one thing you can do that might change the outcome,[2] but it does bear a significant risk of unintended consequences. Just as you realize you've knocked the toast, just as it starts to teeter on the edge, the physics suggests that giving it another good sideways swipe will actually help. The toast will end up on the other side of the room, but because it spends less time pivoting on the edge, it won't be spinning nearly as fast as it falls and may not rotate enough to turn over before it lands. So it stands a decent chance of landing buttered side up – but there is also a decent chance that it will end up under the sofa or stuck to the dog.

Toast starts to spin because it has two things going for it: a point to pivot around, and a force that's pulling the toast around the pivot. It doesn't matter that the force only points straight down and doesn't keep pulling the toast around the circle. What

[1] You may be wondering why the welly in the trebuchet *can* stop rotating and zooms off in a straight line when it is released, but the toast must keep spinning. The difference is that the toast is held together as a single object by internal forces, and as long as it's a single object, it has a fixed amount of angular momentum that must be conserved. If any part of it was released from the rest (maybe a crumb fell off one side), then that piece would travel in a straight line.

[2] That is, only one thing apart from making your toast the size of a matchbox or serving breakfast from a very low coffee table.

matters is that the force is sufficient to move the toast (which it is, if the centre of mass is over air and not table), and that it pulls the toast around the pivot for a little while at least. Once the spin has started, it'll keep going until something stops it.

This is the principle behind the spinning eggs in the introduction. If you think about the many things that spin freely – frisbees, tossed coins, rugby balls, spinning tops – you'll notice that they just keep going. It would be really odd if you flicked a coin so that it span up into the air, and then somehow stopped spinning before you caught it.[1] Anything that spins has angular momentum, which is a measure of its amount of spin. Unless something (such as friction or air resistance) slows it down, the object will spin indefinitely. This is the law of conservation of angular momentum. Something that is spinning will keep spinning, unless something happens to stop it.

I'm pretty sure that when I was a kid, dizziness was considered a sort of internal toy. If you were bored, you could always spin around on the spot, partly to see who could keep going the longest, and partly because it was funny that everyone fell over as soon as they stopped. The spinning itself didn't seem to cause too many problems – the brief and entertaining disorientation comes when you stop. It's a shame that adults don't play this game very often; we might understand ourselves better if we did. The disoriented feeling happens because of something going on inside your ears that you can't see, but your brain is certainly aware of.

Let's go back to the spinning raw and boiled eggs that I was talking about in the introduction. Each egg, still in its shell, is

---

[1] The nice thing about tossing a coin is that it shows that the overall movement of the object and its spin can be independent. The coin would move in the same arc whether or not it was spinning. But if you flick it in the right way, you give it spin as well as upward speed. The spin and the movement of the centre of mass don't interfere with each other.

put down on its side and spun around. After a few seconds of both eggs spinning, you quickly put your finger on the shell of each to stop the rotation. Both eggs stop spinning. You take your fingers away. Then one egg starts revolving again. The egg that is solid has to stop spinning completely when you stop the shell. Both egg and shell have to move together. But when you stop the raw egg, you only stop the shell. The fluid inside is still spinning around; it's not connected to the shell, and so there's no reason for it to stop spinning. So the fluid pushes on the shell until the shell starts to rotate again.

When you spin yourself around, most of you (fortunately) is like the hard-boiled egg. It all has to move together. So when you stop spinning, your brain and nose and ears all stop, too. But not your inner ears. There are small, semicircular canals in each ear that are filled with fluid precisely because that makes them behave like the raw egg. The fluid doesn't have to match the movement of its container because it's not attached to it. This is one of the ways that your body senses where you are; tiny hairs detect how the fluid is moving around, and your brain matches that up with what you see. If you rotate your head, the fluid in the curved canal doesn't rotate as quickly, so it flows around the canals because it hasn't caught up yet. But if you spin for a while, this fluid starts to spin, too. It only takes a few seconds to catch up, and then the fluid in your ears is spinning steadily with the canals, matching the movement of its container. When you stop suddenly, the fluid doesn't stop. Just like the raw egg, the container has stopped, but the fluid keeps going. So your inner ear is telling your brain that you're moving, but your eyes are telling your brain that you're not. That is when you feel dizzy, as your brain tries to work out what's really going on. Eventually, the fluid in your inner ear does stop spinning, because its container has stopped. The dizziness fades away.

This is one of the reasons why pirouetting ballerinas keep facing in one direction as they spin, and then very quickly move their head all the way around to get back to the same direction as their body catches up. With this very quick stop–start motion, the fluid inside doesn't get up to a steady spin, so the ballerina doesn't feel disoriented when she stops.

There are two aspects to the conservation of angular momentum. The first is that something that isn't spinning needs a push to get it going. It can't just start spinning by itself. And the second is that something that is already spinning will keep spinning unless something pushes on it to make it stop. In our everyday lives, it's often friction that provides the push to slow things down. So the spinning top eventually comes to a halt and the spinning coin slows down so much that it falls over. But in situations where there is no friction, things really will keep spinning indefinitely. That's why the Earth has seasons.

In northern England, the seasons are the rhythm that gives all my memories a cosy home. Long walks along the Bridgewater Canal on hot summer days, hockey matches in the autumn drizzle, driving back from Polish Christmas Eve dinner in a frosty chill, the excitement of spring days getting longer – the variety was part of the joy of it all. One of the hardest things about living in California was the absence of that rhythm; it felt as though time wasn't moving, and that was intensely disconcerting. I continue to feel the seasons very strongly today. I like to be able to identify my place in the yearly cycle by the cues that still mark it, even in a modern society: animals, the air, the plants and the sky. And the foundation of all those riches is the bit of physics that keeps things spinning unless something stops them.

Spin has a direction, the axis around which everything is rotating. We imagine the axis of the Earth as a line that goes from the South Pole to the North Pole, sticking out slightly, pointing away into space. But because it's been clonked by solar system debris in the past (especially the huge collision that made the Moon), the spinning top that is the Earth doesn't point straight up relative to the rest of the solar system. Imagine looking down on the solar system, with the Sun in the middle and the planets circling around on a flat plane. The axis of the Earth points slightly off to the left. And now that it's spinning about that tilted axis, it has to stay spinning about that same axis. So when the Earth is on the left of the Sun as we're looking at it, the north end of the axis is pointing away from the Sun, out into space. But six months later, when the Earth is on the right of the sun, the north end of the axis is still pointing to the left – which is now towards the Sun. The spin axis of the Earth does not change direction as it goes around the Sun – there's nothing pushing on it, so it must keep going as it did before. But that means that the North Pole gets more or less sunshine, depending on where the Earth is on its orbit. This is where our seasonal cycle comes from.[1] We have a day/night cycle because the Earth keeps spinning, and a seasonal cycle because the axis of that spin is tilted.[2]

Spinning is part of our lives in lots of ways. But there's one device in particular that relies on spin, and it's one we might all be seeing more of in the future: a flywheel. Anything that's spinning has extra energy because of its spin. So if a rotating object will keep spinning indefinitely, that also means it can act as a store of energy. If you can get the energy back as you slow the spin down, you've effectively got a mechanical battery. This is what a flywheel is, and it isn't new; they've been around for centuries. But a new wave of flywheels is about to arrive in our

[1] The complete gravitational picture is slightly more complicated than this, but the basic idea is correct. Look up Milankovich cycles if you'd like to know more.

[2] Although the Earth has been spinning continuously since its formation, it has slowed down ever so slightly because the tug of the Moon provides a very gentle brake. It's only causing a tiny change, but every one hundred years a day on Earth gets longer by about 1.4 milliseconds. Every few years, a leap second is added to a year to take this into account.

society, a set of very efficient modern devices that could help solve a really thorny problem.

One of the biggest challenges for any energy grid is matching up supply and demand on very short timescales. If everyone cooks dinner at about the same time, energy use across the country will rise for an hour or so and then fall. Ideally, someone monitoring the system would allow energy into the grid as it's needed, to match that spike. But that's a problem if the energy is coming from a coal-fired power station that takes hours to start and stop. And you may not even be in control of the rate or timing of energy generation. One of the difficulties with many renewable energy sources is that you can't dictate when they're generating energy – it's easy to make hay (or store energy) when the sun shines, but what if that's not when you need it?

Surely, you might say, all you need is a battery to store the extra energy until you can use it? But electric batteries aren't really up to the job. They're expensive to manufacture, they're often based on relatively rare metals, they have a limited number of charge/discharge cycles, and there are limits to how quickly they can store and release energy. In response, over the past few years, along have come some prototype flywheel projects. And it looks as though this technology may offer a workable solution, at least some of the time. A flywheel is a heavy spinning disc or cylinder, with bearings that are as frictionless as possible. Once it's spinning, it will keep spinning. And since there is energy associated with rotation, that spin can store energy. You use any excess energy in the grid to get your flywheel spinning, and it will just keep spinning, holding on to the energy. Then, when you want energy back, you slow the wheel down by converting the energy into electricity. There's no limit to the number of times you can charge and discharge the flywheels, and they can release their energy very quickly. You only lose about 10 per cent of the energy you had to start with, and very little maintenance is needed. Even better, you can make them to suit your needs: a small one to go with the solar panels on your roof, or a vast bank of them to moderate spikes in the whole energy grid. Small portable flywheels are even being tried out on hybrid buses, storing energy as the bus brakes and supplying it back to the wheels when the bus needs to speed up again. Flywheels are appealing because they're based on a beautifully simple idea – the conservation of angular momentum. Eggs and spinning tops and stirred tea all follow the same principle. But it takes efficient modern technology to turn it into a practical solution. It's still early days for the new incarnation of this technology, but you may well be seeing spinning flywheels around much more often in the future.

8

When opposites attract

Electromagnetism

A SELF-TIDYING BAG SOUNDS like a pipe dream. But it might not be as impossible as it sounds. One day last year, I'd popped into London's Science Museum to buy some lovely spherical magnets (some for a friend and some for me – that's how it should be with science toys, right?). I stopped off for a hot chocolate and a few minutes' play with my new toys, and then I tucked the clump of magnets into the jumpers at the top of my travel bag and carried on. Two days later, in Cornwall, I remembered that I hadn't seen the magnets for a while and dug around to see where they were. When I found them, they were right at the bottom of the bag and the magnet cluster had expanded to include seven coins, two paperclips and a metal button. I was just congratulating myself on having found a new way of keeping my bag tidy when I noticed that there was plenty more loose change at the bottom of my bag that hadn't joined in this new game. So I started sorting the coins to see which ones stuck and which ones didn't. Some 10p pieces did, and some didn't. Nothing with a value higher than 20p stuck. Most 1p and 2p coins did, but not the ones with dates earlier than 1992.

The thing about magnets is that they're very selective. They have no attraction at all for most materials – plastics, pottery, water, wood or living things. But for iron or nickel or cobalt it's a different story. These will leap towards a magnet if they are free to do so. It's a weird thought, but if iron weren't one of the most common materials in our world, we'd probably never come across magnetism in our everyday lives. Just this one element makes up 35 per cent of the Earth's mass, and steel (which is mostly iron with a few other things mixed in) is an essential part of our modern infrastructure. If refrigerator doors weren't reliably made of steel, fridge magnets would never have happened. But steel is everywhere, so magnetism is common.

The magnets in my bag were sorting the coins according to their composition. Modern 1p and 2p coins have a steel core with a thin layer of copper on the outside. Before 1992, they were 97 per cent copper. The old and new pennies look almost identical to me, but the magnets are responding to their hidden innards.[1] The silvery 20p coin doesn't stick to magnets because, oddly, it's mostly copper. So are the older 10p coins, but any minted since 2012 are nickel-plated steel. Everything sticking to the magnet was mostly iron, even the 'coppers'.

A magnet is surrounded by a magnetic field, something you might call a 'force field'. That means there's a region around it that can push and pull on other objects, even if the magnet itself isn't touching them. That's a bit of a weird idea, but it's the way the world is. The problem with magnetic fields is that we can't see them and can't usually feel them, so they're hard to imagine. But we do see the effect they have, and that can help our imagination along. And the most important thing about magnets is that they all have two distinct ends, a north pole and a south pole.

---

[1] The newer pennies are ever so slightly thicker because they've been made to weigh exactly the same as the older ones (the same mass of steel takes up slightly more space than copper). That's why vending machines have to be changed when the Mint alters the material of the coins – different metals take up a different amount of space for a given mass. Vending machines also test that the magnetic properties are correct for the coin type.

Magnetic north of one magnet will attract magnetic south of another magnet, but two north poles will repel each other. My coins weren't magnetic to start with, but the magnets played a clever trick to attract them. Inside each one of my new 1p coins, different regions of the iron have magnetic fields pointing in different directions. These regions are called domains, and the magnetic fields of the atoms inside each one are all lined up. Each domain has an overall magnetic field of its own, but because all those domains have their magnetic north pointing in randomly different directions, the whole thing cancels out. As I brought a coin close to one of my magnets, the strong magnetic field from the magnet was busily shoving on all the individual domains in the coin. The atoms didn't move, but their magnetic field swung around so that the north end was as far from the north of my magnet as possible. That left all the south poles of the coin domains lined up so that they were closest to the magnet. And since opposite magnetic poles attract, the south pole of the coin was attracted to the north pole of the magnet and the coin stuck. And as soon as I took the coin away from the magnet, all its magnetic domains went back to being randomly oriented.

It's a weird phenomenon, but we humans have learned to make use of it in ways that now permeate our lives. It starts with coins and paperclips and fridge magnets, but ultimately, magnets are essential to the way we generate power for our world. At the heart of every single device that feeds electricity into our power grid, there's a magnet. However, magnets don't do it alone and magnetism is only half of the story. It's linked in a very fundamental way to electricity, something so vital to modern society that we hardly notice it any more.

It was the science-fiction writer Arthur C. Clarke who said that 'any sufficiently advanced technology is indistinguishable from magic'. Electricity and magnetism together are responsible for more magically advanced technology than almost anything else. When you look really hard at the physics, you can see that these invisible forces are two sides of the same phenomenon: electromagnetism. They are bound together, each influencing the other. But before we look at the connection, let's dig a little bit deeper into the side that we're most familiar with: electricity. Unfortunately, the first time most of us experience electricity in a direct way, it hurts.

*

Rhode Island is a tiny, friendly fragment of the American north-east, and it was my home for two years. Its official nick-name is the Ocean State, and the locals have entirely missed the irony of nicknaming the smallest state in the US after the most gigantic feature on the planet. The mentality of Rhode Islanders rests on two pillars: the coastline and the summer. Life there is about sailing, crab shacks, snail salad,[1] and the beach. But the winters were cold. The tourists vanished, the locals hibernated, and the olive oil in my kitchen solidified if I turned the heating off when I went out.

On the best winter days, I woke to a distinctive stillness that told me before I had even opened my eyes that snow had fallen overnight. For someone brought up in grey, damp Manchester, this was always hugely exciting. I loved it all, apart from one sin-gle repeated moment. After pulling on snug winter boots, shovelling the snow from my path and laughing at the squirrels digging in the white stuff, I'd stomp out to my car in the stillness. And every single snowy morning, as I touched the car for the first time, I'd be greeted with the sharp snap of a painful electric shock. I never quite remembered in time. Ow!

---

[1] I'm not even joking. They're very proud of it. Little Miss Vegetarian here was excused, but I gather it's mostly made up of giant marine molluscs and garlic.

It always felt as though it must be the car's fault in some way, but with hindsight, it wasn't the car that was to blame. As I walked down the path, I was carrying a small flock of sneaky passengers looking for an escape route. The pain was just a side-effect of their jumping ship. These passengers were electrons, incredibly tiny fragments of matter and some of the most fundamental building blocks of our world. The wonderful thing about electrons is that you don't need a fancy particle accelerator or a sophisticated experiment to know that they're moving about. In the right situation, our bodies can detect their movement directly. It's just a shame that our bodies register this astonishing detection as pain.

It all starts with what's in an atom. At the core of each one, there's a heavy nucleus that makes up almost all the 'stuff' of the atom. This nucleus has a chunky positive electric charge, so it will almost never be alone. Electric charge is a strange concept, but it holds our world together. There are only three building blocks that make up almost everything we see – protons, electrons and neutrons – and they each have a different electric charge associated with them. Protons are much more massive than electrons, and they have a positive charge. Neutrons are similar in size to protons but have no electric charge. And each electron is minuscule by comparison but has exactly enough negative electric charge to balance out one proton. This mixture of building blocks dictates the structure of our world. In the centre of each atom, protons and neutrons cluster together to form a heavy nucleus. But an atom needs to be electrically balanced. Electric charges affect the world because different charges attract and like charges repel (as we saw with my magnets and coins). So tiny electrons swarm around the massive nucleus because they are negatively charged, and therefore attracted to the positive charge in the centre. Overall the positives and negatives cancel each other out, but the attraction holds the atom together. All the matter that we see is full of electrons, but because everything is balanced we don't notice them. They become noticeable when they move.[1]

The problem is that when you've got tiny, nimble players like electrons in the game, things don't always stay balanced. When two different materials touch, electrons quite often hop from one to the other. It happens all the time, but it doesn't normally matter because the extra ones will usually find a way back quite quickly. Walking around my cottage in socks wasn't a problem – a few electrons were hopping from the nylon carpet to my feet with every step, but they'd soon find their way back. As soon as I pulled on my fleece-lined, rubber-soled boots, things changed a bit. The wandering electrons were hopping from the carpet to the rubber soles, just as before. But, nimble though electrons are, there are some materials they can't easily make their way through: these are called electrical insulators, and rubber is one of them. The rubber has plenty of its own electrons, but it can't easily soak up any extras. As I was packing my bag for the day, finding my coat and tidying up from breakfast, I was accumulating electrons as they quietly hopped on board. This led to extra electrons spreading themselves out around the outside of my body. By the time I stepped outside, I was the vehicle for a few thousand billion extra electrons, a gigantic number but still a minuscule fraction of my body's own electron cohort.

Why didn't they escape? Each one of those extra negatively charged electrons was being repelled by the others – any route away would be better than staying put. But my boots stopped

---

[1] Molecules are formed when electrons shift around so they're being shared between different nuclei: this sharing forces the nuclei to stay close together, forming a single molecule made up of different atoms. The only thing that holds atoms and molecules together is that positive charges attract negative charges. Sometimes, electrons shuffle around between molecules, changing which nuclei are pinned together, and the pattern made by those nuclei. We call that a chemical reaction. Chemistry is the study of this electronic dance, and the fantastic complexity it produces.

them leaving via the floor. There is another common escape route: moist air. Humid air contains lots of water molecules, each with a positive segment that could host an extra electron for a while. Most days, my extra flock of electrons would have escaped one by one, as they hitched a ride with floating water. But cold days after a heavy snowfall are often dry. There is very little water in the air, so the air offered no way out.

And so, every dry, snowy day, I'd walk down the path from the cottage to my car, completely unaware of the billions of negatively charged passengers, at least until their opportunity knocked. My car sat on the ground, a vast reservoir of balanced electrons and nuclei. The split second when my bare fingers first made contact with the metal of the car was like the opening of an escape tunnel. Metal is an electrical conductor, so electrons can flow through it very easily. My electron passengers surged through the skin of my finger tip, finally free when they met the car. The nerve endings in the skin jangled as the mob whooshed past, directly stimulated by the flow of electrons: an electric current. And I would curse, the magic of the snow temporarily forgotten.

These days, an electric shock is the most direct experience of electricity that most of us have. And yet we're surrounded by the stuff. The walls of our buildings, our electronic devices, our cars and lights and clocks and fans are all buzzing with it. But electricity isn't just about plugs and wires, circuits and fuses. Those are just the crude trophies advertising the human manipulation of this phenomenon. Our planet is humming with electricity in lots of surprising places. Even the humble bee is in on the act.

Imagine a warm, peaceful, lazy day in a very English garden, with a chaffinch pecking fussily at the edge of the lawn. Behind him, jaunty ranks of flowers are engaged in a slow but fierce battle for water, nutrients, sunlight and the attention of pollinators. The scent of jasmine and sweet peas is drifting across the grass, advertising their wares. A honeybee buzzes along the flowerbed, inspecting what's on offer. This may look like a relaxed scene, but for the bee, this is hard work and efficiency matters. It's costing her an enormous effort to stay in the air. She has to flap her tiny wings two hundred times every single second, and the constant pummelling of the air is so powerful that it sends out vibrations we can hear: the buzzing. If you're the size of a bee, air resistance is much greater than it is for us, so it's much harder to push past and through all those air molecules. Thumping the air like this is not an elegant way to fly. But it works, and she hovers for a second next to a pink petunia before deciding that this will be her next stop. As she's flying in, but just before she touches the flower, something very odd happens. Pollen grains that were sitting in the centre of the flower suddenly hop across the air gap to the bee's fur. And as she settles on the flower, more pollen settles on her. She hasn't taken a single sip of nectar yet, but she's wearing a coat of the plant's DNA, and it's almost as though it's deliberately jumping on board.

It turns out that flying makes a bee very attractive, quite literally. It's not because of her appearance or behaviour. It's because our bee is electrically charged, although only very slightly. Just like my electric shock, it's because some electrons have shuffled themselves around. But this time, no one's getting hurt.

The bee's own electrons are hovering around the edges of each molecule in the bee's wings. If something is rushing past the bee very quickly (air, for example) and anything is going to get knocked off, it's going to be an electron. And this is what happens. It's the same as rubbing a balloon against a woolly jumper – static electricity builds up, which just means that something has more or fewer electrons than it should. As those frantic wings were shoving air molecules out of the way, electrons from the wings were being rubbed off and were floating out into the air. The flying bee was left with a slight positive charge because it no longer had enough electrons to cancel out the positive charge of all the protons in its atoms. It's small though, certainly not enough to give a human an electric shock.

As the bee approaches the flower, it attracts negatively charged electrons to the surface, and repels the positive charges. Just as the north pole of a magnet tugs its opposite (magnetic south poles) closer to it, so a positively charged bee tugs on negatively charged electrons. When it's very close to but not yet touching the flower, the positive charge of the bee pulls the surface of the pollen hard enough to tug a few grains off the flower, across the gap and onto the bee. Then the pollen sticks to the bee's fur, just like a statically charged balloon sticking to a wall. When the bee flies on to the next flower, that pollen will travel with it. Pollination by bees would work without the static electricity, just because the bee's fur will touch the pollen when the bee lands on the flower, and the pollen will cling to the fur because it's sticky. But the shifting of a few loose electrons so that pollen can jump the gap definitely gives it a boost.[1]

Electrons are tiny and mobile, and so when an electrical charge moves, it's usually the electrons that are providing the transport. They move around quite a lot, but we don't usually notice. Negatively charged electrons will repel each other, so if lots of them accumulate in one place, they'll push each other away and drift apart. A significant charge never builds up. But there are two possible situations that can stop the drifting and trap the charge: either the electrons have nowhere to go, or they can't move. When the bee is flying, the positive charge indeed has nowhere to go, so it builds up on the outside of the bee's body.

---

[1] There's one more twist in the tale of the bee. Researchers from Bristol University discovered in 2013 that each flower has a slight negative charge that's neutralized as the bee arrives. They demonstrated that bees can tell a neutral flower from a negatively charged one without landing on it. They suggested that bees might avoid neutralized flowers, because it suggests that another bee got there first and took the lion's share of the nectar. For more on this, see the papers by Clarke et al. and Corbet et al. listed in the References section at the end of the book.

But it's the other situation, the one where the electrons can't move, that gives us our spectacular control of electricity. If the bee settles on a plastic plant pot, the positive charge can't move into the plastic because plastic is an electrical insulator. That means that even though the plastic has lots of electrons of its own, they're tightly bound to its molecules and they can't move. It's hard to add or subtract a few extra electrons to or from the mix, because they can't sneak in among the others. This is what defines an electrical insulator – it has no capacity to take on or give up a few extra electrons. So when a bee lands on a plastic plant pot, the positive charge stays put on the bee. A metal garden fork would rob the bee of its charge immediately; metals are electrical conductors and electrons can shift about inside them very easily. The reason metal behaves like this is that all its atoms share their outer electrons in a giant surrounding mob. Since these electrons are moving about all the time and none of them belongs to any particular atom, it's easy to add or subtract a few.

Our society can only have and control an electrical grid because we have both types of materials, conductors and insulators. That's all you need: a mosaic of materials that is really just a maze for electrons in which some paths are easier than others, and a way of controlling some parts of that pattern. Once you've got those basics, you have amazing control over the world.

*

Static electricity is a start, but the real power comes when you start to move electrons and electrical charges more systematically. Our electrical grid, the network we use to move energy around, is an astonishing resource. By pushing electrical charge down wires, and controlling it using tiny switches and amplifiers, we can deposit energy wherever we need it. An electrical circuit is just a way of redistributing electrical energy. The most important thing about a circuit is that it is just that – a circuit. It has to be a loop, so that electrons can keep shuffling around it without building up at the far end. Every circuit must begin and end at a power supply, something that will keep the electrons on the move, taking them in at one end, pushing them along and putting them back into the circuit at the other end. The power supply is a bit like a lift that carries people up to the top of a very long slide. The people can go down the slide and up to the top again, round and round all day, as long as there's a lift to give them enough energy to get back to the start. The rule of every circuit is that you have to lose all the extra energy from the power supply before the electrons get back to where they started from.

An electron shuffling along a wire is all well and good, but what's pushing it around the circuit? We've said that the first thing is to have an electrical conductor, something that provides a path down which an electron can move. But the other thing you need is a force to push it with.

A fridge magnet and a balloon charged with static electricity are both weird for the same reason: they show that it's possible to have an invisible force field. That is, one stationary object is pushing or pulling on another one nearby, but you can't see what's doing the pushing. This similarity isn't accidental, but the real link is only obvious once you start moving the electrical or magnetic fields around. First, let's go back to that principle of a force field. It's not just humans that can make use of them.

The stream bed is a murky brown maze of rocks, plants and tree roots. It's dusk and the muddy water is flowing lazily through and over the obstacle course. A metre below the surface, two small antennae are poking out from beneath a pebble, twitching as they taste the water. Something moves nearby and the antennae vanish. This freshwater shrimp is a scavenger, hungry but vulnerable. Upstream, a hunter slides into the dark water. It paddles along the surface towards the centre of the stream with two webbed front feet, then shuts its eyes, closes its nose, seals its ears and dives. The platypus is ready for dinner.

If the shrimp stays perfectly still, it will be safe. The platypus swims quickly, picking its way confidently through the maze even though it's currently blind, deaf and unable to smell anything. Its flat bill sweeps from side to side, scanning the mud. Another foraging shrimp feels the water move as the platypus approaches and snaps its tail, jerking backwards into the gravel. The hunter swerves towards it. The signal forcing the tail muscle of the shrimp to contract was an electric one. That electric pulse created a temporary electric field centred on the shrimp. This electric disturbance flashed through the surrounding water, exerting tiny pushes and pulls on nearby electrons. It lasted for a fraction of a second, but it was enough. A platypus has an array of forty thousand electrosensors on the upper and lower surfaces of its bill. The simultaneous water movement and electric pulse were all it needed to get a direction and range. The bill hammers into the sand in exactly the right place, and the shrimp is no more.

The shrimp's movement condemned it because the act of moving changed its electric field. Every electric charge pulls or pushes on other electric charges around it. An electric field is just a way of describing how strong that push or pull is in different places, while talking about electric signals means that an electric charge moved somewhere, and something nearby noticed the change because the push on it increased or decreased. Since all muscle movements involve moving electric charges around inside the muscles, they all generate electric fields. So electrosensing is a reliable hunting technique underwater if you're close enough to your prey, because no amount of colourful camouflage can disguise an electric signal. Any animal has to move eventually, and the tiniest motion will generate an electric signal that can give it away.

If that's the case, why aren't we more aware of the electric fields that we generate ourselves? It's partly because those fields aren't very strong, but mostly because electric fields decay quickly in air, which doesn't conduct electricity. Stream water (and especially salty ocean water) is a very good conductor of electricity, so electric signals can be detected from much further away. Almost all the species that use electrosensing are aquatic (bees, echidna and cockroaches are the known exceptions).

In an electric circuit, the electrons move because there's an electric field inside the wire. That electric field is pushing on each electron, shoving it along. But where does the electric field come from? A good place to start is with a battery. Batteries come in all shapes and sizes, but there is one set in particular that I will never forget. They were chunky sea batteries, and I worried about them because they were floating free in a giant storm, powering my one shot at an important experiment.

To study the physics of the ocean surface in storms, we need to go and look at that surface. The ocean is such a complicated environment that theorizing from a nice warm office is of limited use unless you're sure that what you're working on is definitely based on reality. But even when you get 'there', on a ship miles from shore in rough seas, it's still difficult to touch the region I'm interested in – the water just a few metres below the sea surface. Knowing what happens there will improve our understanding of how the oceans breathe, and will contribute to better weather forecasts and climate models. But to see the details, you need to be in it; and it's a violent, messy, dangerous place to be. I can't swim in that water, but my experiments have to. The experiments need power, an electricity supply, and they need it while they're bobbing up and down in the waves, free of the ship. You can't plug them in, so you have to rely on batteries. And fortunately for me, electrical circuits work just as well when they're bobbing up and down as they do when on dry land.

*

The bosun scowled at the horizon, stuffed his hands deep into the pockets of his paint-splattered hoodie and swayed along the deck of the ship towards me. It was November in the North Atlantic, and I hadn't seen land in four weeks. Everything was always either going up or going down as we clung to a heaving grey sea that merged with grey sky in every direction. The roll of electrical tape I'd just put down on the deck took advantage of my temporary distraction and skidded across the deck until it met the bosun's boot. His thick, cheery Boston accent seemed comically out of place in this forbidding environment. 'How long you gonna be?'

For me, the worst bit about doing experiments at sea has always been these final checks before letting the experiments float free. I was nervous, and this bit was my responsibility alone. To measure the bubbles just beneath the breaking waves, I was using a large yellow buoy with a variety of measurement devices strapped to it. The bosun was in charge of manoeuvring this beast off the ship and into the rolling sea, but I had to make sure that it was ready. The storm that was coming would be a big one, and I desperately wanted good data from it. 'I'm just about to plug in the batteries, and then I'm ready to go,' I said. The monstrous yellow buoy, 11 metres long, that carried my experiments was strapped to the deck, shackled securely until it was safe to release it. I started with the armoured camera near the top and put my hand on the power connector, following the wire all the way down to the bottom of the buoy where the chunky batteries were and then plugging it in. Then back up to the acoustical resonators. Hand on the power cable, follow it down to the batteries, plug it in. Check that the connection is secure. Check again. Back again to the other camera. These experiments could carry out incredibly delicate and sophisticated manipulation of the physical world, but only when provided with electrical energy. And the providers were four cumbersome lead–acid sea batteries that weighed 40 kg each, and whose basic design hadn't really changed since they were invented in 1859. But they worked.

When it was time, we scientists huddled in our oilskins at the far end of the deck and the crew and crane took over, manoeuvring the swaying monster over the side and into the dark ocean. As the last rope slipped free, there was a weird shift of perspective, and the huge yellow beast became a vulnerable bobbing piece of flotsam, tiny compared to the vast ocean and frequently hidden by the waves. A burst of chatter spread along the ship's rail about how the buoy was sitting in the water, and the speed at which it was drifting away from the ship. But I wasn't thinking about any of that. I was thinking about electrons.

Below the waterline, the dance of the electrons had started. They were shuffling out of the battery, around the circuits carried by the buoy and then back into the other side of the battery. There were a fixed number of electrons held in the circuit, all just going around the same loop. The electrons don't get used up – they just go round and round. The trick is that it takes energy to push them around, and they give that energy away as they travel. The source of that energy is the battery, and a battery is a very ingenious device.

The clever thing about batteries is that they join up a chain of events. Each link in the chain supplies the electrons that the next link needs; and so once a battery is connected to a circuit, everything is in place for electrons to flow around the loop. These sea batteries had two terminals sticking out to connect them to the outside world. Inside, each terminal was connected to one of two sheets of lead, but those two sheets weren't touching. The space in between the lead sheets was full of acid, which is why they're called lead–acid batteries. There are two ways in which the lead can react with the acid. There's one that needs some extra electrons from somewhere, and there's another one that gives away extra electrons. A lead–acid battery is charged when those two reactions have been pushed as far as they can go.

When I plugged the equipment into each battery, I was providing a path all the way from one lead sheet through my experiments to the other lead sheet. And then there was the crucial last piece to the jigsaw: because of the chemistry at the lead plates, there was an electric field down the wire. Electrons were being pushed along the wire, away from one lead sheet towards the other. They couldn't get there across the acid, so the only option was the outside circuit, the long way round. Once the electrons have a path with an electric field pushing them down it, the reactions can undo themselves because the chain is complete. One set of lead plates gives electrons to the acid, and then the acid passes this charge on to the lead at the other plate. The lead there takes electrons as it reacts, and the whole thing keeps going because the electrons can then shuffle around the circuit back to the first set of plates. The really important fact is that on that trip through the camera around the back, the electrons have some extra energy to get rid of. This is electricity. And if you arrange it so that on their way they have to pass through a sophisticated electrical circuit, hey presto: you can put that energy to work, and you've got a useful battery.

As I leaned over the rail of the ship watching that bobbing yellow buoy, I was imagining this dance. The camera would switch on, creating a pathway for electrons from the battery, and they would bounce their way up the stem of the buoy, into the camera housing. You have control over where the electrons go because you know that they'll take the easiest path. So you arrange a path through the maze that is made of conducting material. The power cable is metal, easier for an electron to move through than the plastic coating around the wire, so you know that electricity will flow down the wire rather than escaping into the surrounding material. Beyond that, the most basic element of control is a switch. A closed switch is just a place in the circuit where two parts of electrical wire touch. They're not glued together, but when they're touching, electrons can move between them. To stop the flow, you just move one of the wire ends away from the other. Electrical flow stops because there is no longer an easy route through.

Once inside the camera, the path of the electrons would split, some shuffling into the computer and some into the camera itself. And the thing about electrical circuits is that in the end all roads lead to Rome, or in this case, back to the battery. The massive yellow buoy was just the skeleton for this branching flow of electrons, and the electrons themselves were generating electric and magnetic fields, pushing and pulling on camera shutters, acting as timers, generating bursts of light and recording data in a huge, intricate synchronized sequence, before shuffling back to the battery.

And all of that was happening while the buoy was being shoved around by the huge waves (8–10 metres high in some cases) of an Atlantic storm. We bobbed about on the research ship and waited, living a life where gravity was an uncertain friend and a tenuous grasp on order was maintained only by imprisoning things with Velcro or elastic bungees or rope. After three or four days, the chemical reaction in the battery had finished – it was back in its original uncharged state. There was no more stored energy left, the electrons could not be pushed around the circuits, and the dance was over. The buoy went back to being an inanimate shell of metal and plastic and semiconductors. But the data had been stored in solid state computer memory, and it was safe.

A few days later, when the storm was over, we tracked the buoy down and hauled it back on board. I'm always extremely impressed by the skill of research ship crews at fishing things out of the water. Ships don't move sideways, and they're slow to turn around or change direction. To stand a chance of getting the buoy back, the captain had to bring his 75-metre vessel right alongside it, managing both to avoid running it over and to get close enough for the bosun to reach over and catch it with a long boat-hook. And they usually succeeded on the first try.

Then it was our turn again. The batteries were plugged into the ship's power supply, providing the energy to push the chemical reactions back the other way ready for the next deployment. The experiments were detached and brought inside, with the exception of the camera. This got left outside in the freezing cold, because the dance of the electrons has a downside, and my poor PhD student was about to pay the price for it.

Possibly the most fundamental physical law we know of, one that has been shown to be accurate time and time again and has never been disproved, is that of conservation of energy. It states that energy can never be created or destroyed, but only shifted around from one form to another. The battery had chemical energy, and the chemical reactions converted that to electrical energy, and then somewhere between one terminal of the battery and the other one, that energy moved on. But where did it go? Things happened – the camera took pictures, the computer programs ran and data were recorded. But none of that stored the electrical energy in a new place. The energy just leached away, unnoticed. There is a price to be paid for moving electrons around, and it's the generation of heat. Any electrical resistance inflicts an energy tax on the electrical energy moving through it. Even though the electrons will pick the path of least resistance, some tax must still be paid.[1]

The camera was housed in thick plastic, a material that transmits heat very badly. When the camera was running, all the energy of the electrons was eventually converted to heat as they flowed around the system. That didn't matter in the water, because the ocean where we were was about 8°C and stole the heat away, cooling the housing efficiently. But air isn't up to that task. In the lab, when the computer was running to download

---

[1] This is all that happens in an electric heater in your home. Electrons are forced through a huge resistance, and their electrical energy is converted to heat. Every other energy conversion process is inefficient, because some energy is always lost to heat. But if heat is what you want, you can have 100% efficiency . . . perfect!

the data, the camera kept overheating. We did our best, but the only solution we found was to leave it outside in a bucket of iced water (helpfully, the ship had an ice machine), and so my PhD student had to spend nine or ten hours starting and stopping the downloads to keep the data flowing while preventing the camera cooking itself. Such is the glamour of field science.

This is why laptops, vacuum cleaners and hairdryers heat up as you use them. The electrical energy must go somewhere, and if it's not converted into other kinds of energy, heat is the inevitable end. Hairdryers use this to heat air; their circuits are arranged to dump energy as heat in a very concentrated way. But laptop manufacturers hate heat, because hotter circuits work less efficiently. There is no way of using electrical energy without paying a heat tax.[1]

So the electrons flow because an electric field is pushing on them. A battery doesn't really provide electrons – there are plenty of those in the world. What it does is provide the electric field to move electrons. And if the circuit is complete, this electric field will push electrons around the loop. So far, so simple. But what are all those numbers on plugs and in tiny font on the safety warnings? Perhaps it's best to take the typical British approach to all problems: find the biscuit tin and put the kettle on.

The most important thing about a teabreak is that it involves both tea and a break. Some of my American co-workers never really understood this, and used to bring along work to continue discussing it over tea. But for the British, the act of 'putting the kettle on' signifies a change of pace. I'm going to do it now, and in this case my kettle is an electric one that I simply fill with water and plug into the mains. My mind is allowed to stop working for a bit, while the kettle gets on with its job.

---

[1] For the pedants out there, yes, there are superconductors. But cooling things down to close to absolute zero uses huge amounts of energy and produces huge amounts of heat. So it doesn't really help if you're after energy efficiency.

Pushing down on the switch does one very simple thing. It shifts a bit of metal and thereby slots the last segment of a circuit into place. Now there's a route through the maze of the kettle, a path made entirely of electrical conductors that electrons can easily travel along. This path is now uninterrupted and it runs from one pin of the plug, through the kettle, and back to the second pin of the plug. In this case, the electric field comes not from a battery, but from a plug socket.

A standard three-pin plug has one long pin at the top. That's called the ground pin. It's completely separate from the rest of the circuit. Effectively, it's doing the job that my car did on those cold snowy mornings – it's there to provide an escape route if any electrons start to build up in the wrong place (say, on the outside of the kettle). So that's not part of the path that's going to power the kettle.

The other two pins, the smaller ones, are going to do the electron-pushing. One of them behaves like a fixed positive charge, and one like a fixed negative charge. As I press down on the switch, I connect up a path that now has an electric field running along it. Electrons along that path are feeling a push away from the negative side and a pull towards the positive side. So as I find the teapot and dig out teabags, the electrons start to shuffle. They're jiggling around quite a bit anyway, but now they have a slight tendency to drift down the wire. And what that means is that overall, there's a movement of electric charge from one pin of the plug, through the kettle, and out through the other pin of the plug.

On the bottom of my kettle, a label tells me that it's designed to work at 230 volts (230V). The voltage is related to the strength of the electric field that's pushing electrons along the circuit. The stronger the electric field, the more energy each electron has to get rid of along the way. That's what a high voltage is telling you – it's saying that this is the amount of energy available for use along the path of the circuit. In terms of the slide analogy from earlier on, the voltage is the height of the slide that the electrons have to shoot down before getting back to the other pin of the plug. The higher the voltage, the more energy each electron has to dump on the way.

I've swilled out the teapot and put the teabags in it, the milk and a mug are out and ready. Now I'm just waiting for the water to heat up. It only takes a couple of minutes, but when I'm thirsty, I'm very impatient. Hurry up! I know what the voltage of the electrical supply is, but that's only part of the story. The higher the voltage, the more energy each electron can give up. But that doesn't say anything about how many electrons are passing through. The fastest way to dump lots of energy in the water is to make sure that lots of electrons are flowing around the circuit. That's what an electrical current is, and we measure it in amps. The higher the current, the more electrons are moving past one point in the wire in any one second. When you multiply the voltage of the supply by the current (in amps) flowing through the circuit, you get the total amount of energy deposited per second. My kettle runs from a 230V supply, and can draw a current of 13 amps, and 230 × 13 = 3,000 (approximately). The base of the kettle agrees – it says that the kettle power is 3,000 watts (3,000W), which equates to 3,000 joules of energy released per second. That's enough to heat my water to boiling in just less than two minutes, but it will lose a bit of heat to the surroundings, so in practice it takes closer to two and a half minutes.

I've no intention of testing this out while I'm waiting for my tea, but they say 'volts jolt, current kills'. The voltage difference between me and my car on that snowy day in Rhode Island was probably 20,000 volts. But only a tiny amount of electrical charge went anywhere, so it didn't do me too much harm. The current was tiny and very little energy was transferred. If I connected up the path between the two plug terminals with my fingers, so that my body took the place of the kettle, it would be a different story.

A high current means that there are lots of electrons, each carrying the same amount of energy. The total amount of energy is huge, because so many electrons are rushing through. It would be far more dangerous than the shock from the car, even though the voltage difference across the pins of the kettle is only about a hundredth of the voltage difference between me and my car. It's the current that matters most when it comes to potential harm to you.

As the electrons shuffle through the metal of the heating element, they're being pushed by the electric field. That makes them speed up slightly, but the conductor is made up of lots of atoms, and so these sped-up electrons inevitably bump into things. When they bump, they lose energy, heating up whatever they bump into. And so forcing lots of charge to move means that there's lots of bumping and lots of heating. That's all the kettle is doing – speeding up electrons so that they bump into things and pass on their energy as heat. The electrons themselves don't travel very far at all – they might drift at about 1 mm per second. But it's enough.

Boiling water has loads of extra energy, and it's amazing that it gets there just from minuscule electrons shuffling about and bumping into things. Amazing, yet undeniable; my tea is ready, heated by electric fields pushing on electrons in a conductor. This is the simplest possible use for electrical energy: converting it directly into heat. But once people had worked out how to build circuits and power supplies and batteries, things got much more sophisticated very quickly.

There is a fundamental difference between the electron dance generated by batteries (any batteries) and what happens when you plug a device into the mains. In any device powered by a battery, the electrons are always flowing in one direction only. This is called direct current, or DC. A standard AA battery will supply about 1.5 volts DC. But the mains current is different – it's alternating current, or AC. That means it switches direction about a hundred times a second.[1] It turns out to be more efficient if you run your electricity supply like this.

You can switch between DC and AC, but it's a bit of a nuisance. Anyone who carries around a laptop power cable will be familiar with this kind of nuisance – it's the small heavy block that sits in the middle of the cable. It's called an AC/DC adaptor, and its job is to convert the AC current from the mains into the kind of DC current that your laptop wants (which is what the laptop battery provides directly). To do that, it needs coils of wire and a bit of circuitry, and it's still tricky to make all the necessary bits any smaller.[2] So for the time being, we're stuck with carrying around the adaptors.

We take electronics for granted today. But in its early days, it was a capricious and uncertain beast. My own grandfather was getting involved in that world just as all its new sophistication was making its way into our homes.

My grandfather, Jack, was one of the first television engineers. Back in his day, electronics could be loud and hot and was certainly capable of generating quite a pong – as my grandmother readily recalls. Her description of the sort of problem that he used to have to fix reminded me of that physicality about early electronics that's easy to forget in these days of smartphones and wifi on tap – and also surprised me with her familiarity with all the components and processes. I'd never really heard her talk about anything technical in my life, and yet when it came to these old TVs, she was comfortable with specialized electrical

[1] So it gets back to where it started 50 times each second – this is what it means to say that the UK mains runs at 50 Hz.

[2] For those who like the details: there are three steps to what the adaptor does. It changes the voltage from 230V to 20V or whatever the laptop needs. And then it has to cut off half of every cycle so that it only gets the current when it's going in one direction and not when it's going back again. After that it smoothes it out a bit to make the same sort of steady current that you get from a battery.

terms I'd never come across. 'Well,' she told me one day, 'one important component was a line-output transformer, and when that went on the TV, it sometimes went with a bit of a bang but it also produced a burning sensation and a smell.' Her northern accent reminds me that this is almost certainly flat-capped understatement. Electrons have always been invisible, but from the 1940s to the 1970s you could definitely tell they were up to something. There was always the risk of a bang or a pop or a hiss, the sudden appearance of a sooty burnt patch or a flash of light that told you lots of energy had just been shunted somewhere it shouldn't have gone. Jack found himself in at the beginning of the new world of television, part of the only generation that had a real feel for the electrical world. By the end of his career, transistors and computer chips had hidden it all away. The tiny exterior of these components conceals a vast and sophisticated interior, incomprehensible from the outside. But before they came along, there were a few decades when you could almost see the magic at work.

In 1935, at the age of sixteen, Jack had started a trade apprenticeship with Metropolitan Vickers, locally known as MetroVick. This giant of the heavy electronics engineering world was based in Trafford Park near Manchester, turning out world-class generators, steam turbines and other large-scale electronics. When he finished his apprenticeship in electrical engineering at the age of twenty-one, he was classed as having a reserved occupation, deemed too useful to go to war, so he spent five years testing aeroplane gun electronics at MetroVick. The first test of these systems was called 'flashing'. You put 2,000 volts across it and if it didn't go bang, it passed. This was the raw end of the taming of the electron, the early stages of wrangling it into submission.

After the war, EMI was looking for people with electronics experience, because early televisions were skittish, complex beasts, needing an expert to set them up and frequent adjustment throughout their lifetimes. So EMI sent Jack to London to train as a television engineer. The tools of this trade were valves and resistors and wires and magnets, the components that could coax electrons to do your bidding. This visually beautiful pot-pourri of glass and ceramic and metal could be made to do something that sounds very simple, something that was at the heart of every television set until the 1990s. It could make a beam of electrons and then bend it; and if you do that right, you can make moving pictures.

Jack learned about 'CRT' televisions, and I love that name because it connects us to the world that existed before electrons had even been discovered. CRT stands for cathode ray tube, and cathode rays were distinctly odd when they were first dis-covered. Imagine the early German physicist Johan Hittorf, in 1867, looking at his latest creation. In the gloomy lab, there's a glass tube with two bits of metal sticking into the space inside at either end, and all the air inside the tube has been removed. This sounds fairly mundane. But imagine how odd it must have been to discover that if you connect a large battery to the two bits of metal, mysterious invisible stuff flowed from one end of the tube to the other. He could tell it was there, because it made the far end of the tube glow, and he could make shadows by putting things in the way. Even though no one knew what was flowing, it needed a name, so it became known as cathode rays. The cath-ode is the terminal attached to the negative end of a battery, and that's where the strange stuff was coming from.

It was another thirty years before J. J. Thomson discovered that what was flowing wasn't really rays at all, but a stream of individual negatively charged particles – the particles we now call electrons. By then it was too late to change the name of the apparatus, though, and it was still always called a cathode ray tube. We know today that applying a voltage across it generates an electric field stretching from one end to the other, and so elec-trons will hop off the negative end and rush towards the positive end. Any particle with an electric charge will be accelerated by the electric field, which means it will be constantly pushed along. So the electrons don't just move towards the positive end because they're attracted by it, they're speeding up as they go. The higher the voltage difference between the two ends, the faster they're going when they get to the other side. In a CRT TV, they can be going at several kilometres per second by the time they hit the screen. That's a significant fraction of the speed of light, the fastest that anything in the universe can travel.

So the same basic process that led to the discovery of the electron in the first place was in use inside every TV in the world until a couple of decades ago. Each CRT TV has a device at the back that produces electrons. The middle of the TV is a completely empty chamber – a vacuum with no air in it – so there are no obstacles at all, and so the electrons 'fired' from the 'electron gun' stream across that empty space until they hit the screen. It's the purest form of electrical current – charged particles moving in a straight line.

*

My aunt opens a box full of bits and pieces she saved from Jack's workshop when he died. There are glass tubes that look like cylindrical lightbulbs, with a weird metallic insect-like structure inside each one. These are valves, used to control the flow of electrons in circuits. Early on, most of Jack's job seemed to involve working out which one of these had malfunctioned, and replacing it. My mother, my aunt and my grandmother clearly have a lot of affection for these, because there were so many of them around back then, and so many different types. And then in the corner of the box there's a large circular magnet, now broken in half.

This is the great connection, and it represents the moment the penny dropped for the physicists of the late 1800s. If you want to control electricity, you need magnets. If you want to control magnets, you need electricity. Electricity and magnetism are part of the same phenomenon. Both an electric field and a magnetic field can push on a moving electron. But the result of the push is different. An electric field will push an electron in the direction of the field. A magnetic field will push a moving electron sideways.

Creating a beam of electrons is all well and good, but the real cleverness of these old television sets was to control what that beam was pointed at. And the deep connection between electricity and magnetism is at the heart of it. As an electron zooms through a magnetic field, it gets pushed to one side. The stronger the magnetic field, the more it gets pushed. So by changing the magnetic fields inside an old TV, the electron beam could be pushed and pulled to a point wherever it was needed. The large, permanent magnet that my aunt showed me was used very close to the electron gun, to do the basic focusing. But the steering electromagnets positioned a bit closer to the screen were being controlled directly by the signal from the aerial. They pushed the electron beam so that it scanned horizontally across the screen, one line at a time. The beam itself was being switched on and off during each line, and where it hit the screen, a bright spot was created. The 'line-output transformer' that Nana mentioned was the bit of kit that controlled the scanning. To make a smooth picture, 405 lines were scanned, 50 times each second, with the electron beam flicking on and off at exactly the right time for each pixel.

This is an incredibly intricate electronic dance. To see any picture at all as a result of it requires a lot of fiddly components all doing the right thing at the right time. So early televisions had lots of knobs and dials to make adjustments, and it sounds as though the temptation to mess with them was too much for many TV owners. Jack had the knack of knowing how to readjust them. It must have seemed like magic at the time. For centuries, craftsmen had been respected for what they could produce, and everyone could appreciate what they'd done even if they couldn't do it for themselves. Now the world had changed. Electronics engineers could make a device function, but it was impossible to see what exactly they'd done or why it had worked.

It's odd that silent, invisible electrons locked away in a vacuum were the key to the huge richness of visual broadcasting with all its sound and spectacle. And for fifty years televisions were based on the same simple principle. Put an electron in an electric field, and you'll speed it up or slow it down. Put that moving electron in a magnetic field, and its path will curve. Leave it there long enough, and it will go round in circles.

The massive physics experiment at CERN in Geneva, famous for the discovery of the Higgs boson in 2012,[1] works on exactly the same principles as a cathode ray tube, although the particles it can shift around aren't just electrons. Any charged particle can be accelerated by an electric field and have its path curved by a magnetic field. The Large Hadron Collider, the experiment that finally confirmed the existence of the Higgs boson, had protons zooming around in its guts. In this case, the speeds reached were incredibly close to the speed of light, so fast that even with extremely powerful magnets to steer the zooming particles, the circle had to be 27 km in circumference.

So the basic setup used both to discover the electron itself and to run the Large Hadron Collider at CERN, a controlled stream of charged particles in a vacuum, also sat in a corner of many homes until very recently. These days, the bulky CRT televisions

---

[1] This discovery caused enormous excitement. Physicists had detected a pattern in the particles that make up our universe, a pattern they call 'the Standard Model' of particle physics. But the pattern could only be correct if one very specific particle existed: the Higgs boson. It took decades to find it, and it was a tremendous boost to confidence in our understanding of our world when it was proven to exist.

have been replaced almost entirely by flat screens. In 2008, sales of flat panel displays overtook CRT screens worldwide, and the world has never looked back. The switch made laptops and smartphones possible because it made them portable. The new displays are also controlled by electrons, but in a much more sophisticated way. The screen is split into many tiny boxes called pixels, and electronic control of each pixel determines whether or not it gives out light. If you have a screen resolution that is 1280 × 800 pixels, that means that you're looking at a grid made up of just over a million individual dots of colour, each separately switched on and off with tiny voltages, and each updated at least sixty times every second. It's an astonishing feat of coordination, but it's still trivial compared to what your laptop gets up to.

Let's return to the magnets. A magnetic field can push electrons around, and so it can control electric currents. But that's not the limit of the interlinking of electricity and magnetism. Electric currents also make their own magnetic fields.

*

As we saw in chapter 5, toasters heat toast very efficiently using infrared light. But the real brilliance of a toaster isn't that it provides lots of heat – your grill can do that; it's that it knows when to stop. The universal rule of toasters is that the bread only disappears down into the innards of the toaster when you press on a lever at one side. If you don't press it all the way down, it pops right back up. But if you push the lever all the way to the bottom, there's a click and it stays put until it's time for the hot toast to pop out of its mini-furnace. I don't need to stand over it, checking how well browned the bread is. When the bread has turned into toast, there will be another mechanical click and the toast will pop up by itself. So as I wander around the kitchen, looking for butter and jam, something is holding the bread in place.

There's a beautiful simplicity about what the toaster is up to. When you put the bread in place, it rests on a spring-loaded tray. The springs underneath are pushing the bread up to its 'popped' position, high above the heating elements. But you're strong enough to push the bread down in spite of those springs. And once the tray reaches the bottom of the toaster, a protruding bit of metal fills the gap in not one, but two circuits. One of those circuits deals with the heating, so electricity starts to flow around the toaster to heat the bread.

But the other circuit is a lot more interesting. The electrons in that circuit shuffle along and around a section of wire that is wrapped around a small lump of iron. It's a bit like a helter-skelter for electrons – they spiral around and around the iron and then out and along the rest of the circuit back to the plug socket. That's all. But because magnetism and electricity are so deeply inter-twined, when an electric current runs through a wire, it creates a magnetic field around that wire. Sending electrons around a coil of wire means that each time the electrons loop around, they're adding to the same magnetic field. The iron core in the middle of the coil reinforces the magnetic field and makes it even stronger. This is an electromagnet. When an electric cur-rent is running through the wire, it's a magnet. When the current stops, the magnetic field goes away. So when you push down the lever on the toaster, you're switching on a magnetic field at the base of the toaster that wasn't there before. Since the bottom of the bread tray is made of iron, it sticks to the magnet. In other words, while I'm poking about in the fridge, a temporary mag-netic field is holding the bread tray in place. The toaster has a timer on the side, and the clock starts when the circuits are con-nected. When the time is up, the timer cuts the power to the whole toaster. Since there's no power to the electromagnet, it stops being a magnet. Nothing is holding the bread down any more, so the springs pop it up.

I sometimes forget that I've unplugged the toaster, but I find out pretty quickly. If I try to push the lever down, it pops straight back up, even if I push it down all the way. That's because there's no power to the electromagnet, so it can't hold the bread tray down. It's such a simple system, and stunningly elegant. Every time you make toast, you're taking advantage of this very fundamental connection between electricity and magnetism.

Electromagnets are very common because it's really useful to be able to switch magnets on and off. They're in loudspeakers and electronic door locks and computer disk drives. They must be continually powered, otherwise the magnetic field vanishes. The kind of magnets you stick on your fridge are called permanent magnets – you can't turn them on or off, or change the magnetism, but they don't need any power. Electromagnets do exactly the same job as a fridge magnet when they're turned on, but they can conveniently be turned off just by stopping the current.

We're surrounded by small, local magnetic fields, some permanent and some temporary. They are almost all made by humans, either to do a useful job, or as a byproduct of something that's doing a useful job. The magnetic fields don't reach very far, and so they're only detectable very close to the magnet. But these are just tiny local glitches in a much bigger magnetic field that stretches around our planet, and this one is entirely natural. We can't feel it, but we use it all the time.

*

It's easy to take a compass for granted, especially if you do a lot of walking, when it's very handy to be accompanied by a needle that always points north. But imagine getting ten compasses, or twenty, or two hundred. You spread them out across the floor, they all point north, and suddenly you see that this isn't just something that happens when you get a compass out. It's there all the time, and it's consistent. You can take your compass collection anywhere on the globe, unpack it, set it out, and all the compasses will swing round and agree on where north is. The Earth's magnetic field is always there, flowing through cities, deserts, forests and mountain ranges. We live inside it, and although we never feel it, a compass will always remind us that it's there.

A compass is a brilliantly simple measuring device. The needle is a magnet, and so one end of it behaves very differently from the other end. Unhelpfully, these two ends of the magnet are called the north and south poles, but it's just a way of saying that one behaves like the magnetic north pole of the Earth, and the other behaves like the magnetic south pole. If you take two magnets and move them about near to each other, you'll see very quickly that it's very hard to push the two north poles together, but that a north pole and a south pole will attract each other very strongly. This is why it's easy to detect the direction of a magnetic field; if you put a small mobile magnet inside a magnetic field, it will spin around until its north and south ends are aligned with the field. And that's all a compass is: a mobile magnet that gives away the direction of any magnetic field you bring it into. We can't see the vast magnetic field of the Earth, but we can see the compass needle respond to it. It's not just the Earth's field that compasses sense, either. Take a compass around your home, and you'll detect for yourself the magnetic fields that surround plug sockets, steel pans, electronics, fridge magnets and even any iron that's been close to a magnet recently.

Compasses, obviously, are mostly used for navigation. Finding your way about on the surface of a sphere is always going to be tricky, but the Earth's magnetic field has provided a fabulously reliable tool for explorers for centuries. The Earth has a magnetic north pole and a magnetic south pole, and anyone with a compass can orient themselves towards one or the other. As a navigational tool magnetism is straightforward, it's cheap, and it never runs out. However, it does have a few caveats attached to it. Caveat number one sounds unexpectedly serious: the magnetic poles aren't fixed in one place. They wander, and they can travel a very long way.

On the day I'm typing this, the magnetic north pole is in the far north of Canada, about 270 miles from 'true north', which is the real North Pole defined by the Earth's spin axis. Since this time last year, the magnetic north pole has moved 26 miles – it's on its way across the Arctic Ocean towards Russia. This sounds spectacularly unhelpful for navigators, although since the world is a big place it's not as bad as it seems. But the magnetic field moves because of where it comes from, and it's a reminder that the innards of our planet are more than just a static ball of rock.

Deep down beneath our feet, the iron-rich outer core of the Earth is churning slowly. It's shifting heat from the centre out towards the surface, and the rotation of the planet forces the molten rock to rotate, too. Because of the iron, the sluggish outer core is an electrical conductor, and that means it can behave like the electromagnet in the toaster. It's thought that the currents running through the Earth's outer core as it turns are responsible for generating our planet's magnetic field. The process is based on the slow shifting of molten rock; and because the details of the rock movements change with time, the magnetic poles meander. They stay approximately aligned with the spin axis of the Earth because the rotation of the iron-rich rock is caused by the rotation of the whole planet, but the alignment is only approximate.

So if you really care about accurate navigation, you need to correct for the current position of the magnetic pole, because it's not the same as the true North Pole. Today's maps show the direction of both poles. I just had a look at an Ordnance Survey map of part of the south coast of the UK, and both magnetic and spin north are marked at the top. I can see that if you

247

followed a compass directly north for 40 miles, you'd end up about 1 mile west of the line towards true north. A map seems like such a permanent record, and yet the magnetic field that you may use to help you navigate with it is fickle. Modern technology means that you and I won't often get lost because of this. But the aviation industry, with one of the most sophisticated modern navigation systems humans have developed, certainly pays attention. For a start, it has to keep relabelling its runways.

Next time you're at or near an airport, take a look at the large signs at the start of each runway. Every runway around the world is labelled by a number, which is its direction in degrees from north, divided by ten. So the runway at Glasgow Prestwick airport was given the number 12, because a plane landing on it will fly in on what's called a 'heading' of 120 degrees. Each runway will have a specific designation that's a number between 01 and 36.[1] But this heading is relative to *magnetic* north, because that's what a compass is telling you. So in 2013, runway 12 at Glasgow became runway 13, to keep up with the movement of the magnetic pole. The runway hadn't moved, but the Earth's magnetic field had. Aviation authorities keep an eye on it all, and correct the runway designations when it becomes necessary. Since the poles move relatively slowly, the changes are manageable.

But the meandering of the poles is only the start. The Earth's fickle magnetic field has far more to offer than navigational assistance. And the clues it leaves behind have provided the final confirmation of one of the most controversial, simple and profound ideas that geologists have ever had. The continents,

---

[1] Or possibly two numbers that are different by 18 (09-27, for example). That's because you can take off and land either way along the runway, but obviously your heading would be different by 180 degrees.

the immense rocky masses that dominate Earth's surface, are moving.

*

In the 1950s, human civilization was whooshing into a new technological and scientific era. The foundations of our modern society were being laid: microwave ovens, Lego, Velcro and the bikini had all arrived and were working their way into popular use. Humanity was coming to terms with the arrival of the atomic age, social rules were being completely rewritten and credit cards had just been invented. And yet, in the midst of all this galloping modernity, we couldn't make sense of the planet we were living on. Geologists had been fantastic at cataloguing the Earth's rocks, but they couldn't explain the Earth itself. Where did all these mountains come from? Why is this volcano here? Why are some rocks so old and some so new? Why are the rocks different everywhere you look?

One of the many observations crying out for a satisfactory explanation was that the east coast of South America and the west coast of Africa looked as though they had once fitted together like jigsaw pieces. The rocks matched, the shapes matched and the fossils matched. How could all that possibly be coincidence? But most scientists just saw it as an unimportant curiosity; it was almost unthinkable that anything that big could go anywhere. In the early 1900s, a German researcher named Alfred Wegener had finally gathered together all the evidence and proposed the idea of 'continental drift'. Wegener suggested that South America and Africa had once been connected, and that one of these huge land masses had broken away from the other and drifted across the face of the planet. Very few scientists took this idea seriously, because the notion of something as gigantic as a continent just drifting 3,000 miles to the west seemed ludicrous. If that was true, what was doing the pushing? Wegener himself suggested that the continents ploughed through the oceanic rock, but couldn't provide any evidence. There was no 'how' and no 'why', and the theory was quickly shelved. No one else had any better ideas, and the question was left alone.

By the 1950s, there still weren't any better ideas around, but there were some new measurements. The lava spewed out by volcanos had iron-rich compounds in it, and it was discovered that each speck of one of those compounds could act like a compass needle, twisting around to line up with the local magnetic field. The really useful part was that when the lava cooled down and formed solid rock, the tiny iron minerals couldn't move any more, so they were locked in position. These tiny frozen compasses meant that a record of the Earth's magnetic field was built into the rock at the moment it formed. When geologists used this record to look at the changes in the magnetic field through the ages, something even more curious came to light. The direction of the Earth's magnetic field seemed to reverse every few hundred thousand years. It completely flipped, so that south became north and north became south. It didn't seem to matter too much, but it was very odd.

Then the geologists got to the sea floor. One of the many unexplained phenomena of the Earth's structure was that several oceans had ridges of underwater mountains running in vast lines across the flat plains of the ocean floor. No one knew what they were doing there. The most famous is the mid-Atlantic ridge, a line of volcanoes that starts above water (the country we call Iceland is just the protruding end of this ridge) and then disappears underwater, where it zigzags all the way down the centre of the Atlantic Ocean almost to Antarctica. Then, in 1960, magnetic measurements showed that the magnetism of the rocks surrounding that ridge was very strange indeed. It was *striped*, and the stripes ran parallel to the ridge. As you went away from the central ridge, the sea-floor rocks had magnetism that pointed north, then south, then north again, and these stripes ran the length of the mountain range. And it got weirder. If you looked at the other side of the ridge, the magnetic stripes there were an exact mirror image.

In 1962 two British scientists, Drummond Hoyle Matthews and Fred Vine, made the link.[1] With hindsight, you can almost hear the solid 'click' as all the strange pieces of geology dropped into place. What if, they said, the sea-floor volcanoes are building new sea floor as the continents move apart? The magnetism at the ridge lines up with today's magnetic field. But as the continents move apart, that rock from the ridges is carried out to both sides of the volcanoes and new rock is made. When the Earth's magnetic field reverses, the magnetism of the new lava will also reverse, starting a new stripe that points in the opposite direction. The reason the stripes are a mirror image of each other is that each stripe represents a period of one magnetic alignment, before it flips back the other way. Other discoveries around the same time showed the places where old sea floor was being destroyed, which was important because the planet itself stays the same size. On the other side of South America, the Andes mountain range exists because that's where old sea floor from the Pacific is being pushed underneath the continent, back down into the Earth's mantle. Once you know that the continents can be shunted around, colliding and separating, creating and destroying sea floor as they go, the patterns of geology make sense. This was the seminal moment in geology, the discovery of plate tectonics. Plate tectonics is now the backbone of everything we understand about why the Earth is the way it is.

So the continents do drift, but they don't plough through the sea floor. They float on top of what's underneath, pushed around by convection currents beneath the Earth's surface. And this

[1] The Canadian Lawrence Morley had also proposed the same idea at the same time, but his paper had been rejected by the journal for being laughably silly.

process isn't just something of the past. The Atlantic Ocean is still getting wider today, by about 2.5 cm per year.[1] Today's magnetic stripe is still being built. It took an astonishing bit of evidence to convince scientists that the surface of the Earth could possibly be so mobile, but the sea-floor magnetism patterns make it undeniable. Today we can measure the movement of all the continents using very accurate GPS data, and we can see the engine in action. But the key to the Earth's past history and present shape was in the magnetism that can be locked into the planet's rocks for millennia.

Electricity and magnetism together form a partnership that is incredibly important to us. Our own nervous system uses electricity to send signals around our bodies, our civilization is powered by electricity, and magnetism lets us store information and marshal the tiny electrons that get things done. So it's striking that our civilization has done so well at keeping the world of electromagnetism under wraps. We rarely experience electric shocks or power cuts, and we're so good at shielding ourselves from magnetic and electric fields that we could live life hardly knowing that they're there. It's both an amazing endorsement of our control of electromagnetism, and extremely sad because we are hiding this extraordinary part of the world away from ourselves. But maybe the future will hold some extra reminders, and we won't forget it completely. As our civilization faces up to its addiction to fossil fuels, one way out seems to be becoming more likely. Power generation won't just happen in remote power stations. Renewable energy can be generated much closer to home, and maybe in the future, we'll see more of where our electrical energy comes from. The face of my watch is a solar panel, and the watch has been running continuously for seven years now. Technologies already exist that will harvest solar energy from our windows, kinetic energy from our footsteps and wave

---

[1] It's often said that it's growing at about the same rate as your fingernails.

energy from our estuaries. And the principles that they're based on are just the principles of electromagnetism.

*

There is one last piece to the electromagnetic pattern. We saw that an electric current could generate a magnetic field in the toaster. But that process also works the other way round. When you move a magnet near a wire, it pushes on charged particles like electrons, and that means that you can create an electrical current that wasn't there before. This isn't just relevant for the future; this is what makes our electricity grid possible now. We can only get energy into our electrical grid by moving magnets around, whether by using turbines in gas-fired or nuclear power stations, or by turning the handle on a wind-up radio. One of the most beautiful and simplest examples of using electricity and magnets to power our world is the wind turbine.

A wind turbine looks serene from the ground, a soaring white strut supporting elegant twirling blades. But the peace is broken the moment you step inside the base of the tower. The innards are filled with a deep, loud hum, and you realize you've stepped into the belly of a giant musical instrument. The one I walked into, in Swaffham in the east of England, is one of very few that have regular visiting hours, and it's more than a little way off the beaten track. But it is absolutely worth the trip.

As you climb the spiral staircase inside the tower, you climb through the hum as it rises and falls. You can feel the structure being buffeted by the wind. You know that you're getting close to the top when the light starts to flash – natural sunlight is being cut out as the blades rotate. And then out you pop, into an enclosed 360° viewing gallery at a height of 67 metres, right underneath the turbine hub. Any sense of serenity is now long gone. The three gigantic blades, each 30 metres long, are whoosh-ing around with considerable oomph and there is no doubt that

253

there is energy up here to be harvested. As the wind rises and falls, the whine and the blade speed respond almost instantly. This alone is impressive enough.

But the point of it all is hidden in the white snout, the part of the mechanism that's just behind the blades. If I push my nose up against the glass and look up, I can see the whole hub rotating. Right above my head, the edge of the hub closest to the tower spins smoothly around a stationary inner ring. This edge is lined with strong permanent magnets, so the magnets are rotating past the inside of the hub. And the inner ring is lined with copper coils, each connected to the circuitry behind. As each magnet whizzes past each coil, it generates a current through the wire. Electrons are shoved through the coil, and then pulled back again by every magnet that goes past. Without the magnets and the wires touching, energy is transferred from the rotation to become electrical energy in the wires. The blades are driving the magnets past the coils, and the rules of electromagnetic induction are creating a current in each coil. This is how electricity is born.

The same principle operates in all our power stations – whether they're producing coal, gas, nuclear or wave energy. Magnets are pushed past wires, and so movement energy is transferred into electrical current. The beauty of a wind turbine is that this is as raw as it gets; the wind turns magnets which generate current. In a coal-fired plant, water is heated to turn a steam turbine, which turns magnets. The outcome is the same, but it takes a few extra stages to get there. Every time you plug anything in, you're using energy that flowed into the grid as a magnet pushed on the electrons in a coiled copper wire. Electricity and magnets are inseparable. Our civilization relies on energy that is harvested and distributed using the dance between these two twins. We have done spectacularly well at hiding the dance away, trapping it in shielded wires and behind walls and in buried cables. We've done so well at masking it that a child born today may never directly see or experience electricity or magnetism at all. Future generations could be deprived of any contact with the elegance and importance of electromagnetism, as the invisible cloak of progress settles over it all. But it matters, because these days the fabric of our civilization is sewn together with electromagnetic threads.

9

A sense of perspective

EACH OF US RELIES on three life-support systems: the human body, planet Earth and our civilization. The parallels between the three systems are powerful, because they all exist in the same physical framework. Having a better understanding of all three may be the best thing that we can do to keep ourselves alive and keep our society thriving. Nothing could be both more pragmatic and more fascinating. So this final section of the book will provide some perspective, taking each of our life-support systems in turn.

Human

I'm breathing, and so are you. Our bodies need to take oxygen molecules from the air and send carbon dioxide back out. Each one of us walks around in our own personal life-support system, a body with an inside and an outside. Our insides can do all sorts of things, but only when supplied with stuff from the outside: energy, water and the right molecular building blocks. Breathing is just one of the supply routes. It's ingenious: expand your ribcage, increase the volume of your lungs, and the mob of tiny jostling air molecules close to your mouth is pushed down your windpipe by the air further away. Take a deeper breath and your chest expands even more, making space for more of the atmosphere to rush in and touch the smallest structures in your lungs. Then, as you relax the muscles around your ribcage, your ribs are pulled downwards by the Earth, pushing the air molecules in your lungs closer together until they jostle each other back to rejoin the outside. Oxygen isn't the only molecule pulled into your lungs that your body can use. As the air passes by the sensors inside the top of your nose, some of the billions of molecules bumping into the walls will happen to collide with a larger molecule attached to the wall, one they temporarily fit into like a key into a lock. The underlying cell senses the molecular click as they fit together. That's the start of our sense of smell – a few floating molecules of the right type bumping into the right place. Inside now has some information about what's outside.

A human body is a vast coordinated collection of cells, about 37 trillion of them last time anyone tried to count, each one a tiny factory. Every single cell needs supplies, but it also needs a safe environment, with the right temperature, pH and moisture level. As you walk through the world, your body is constantly adjusting to adapt to the conditions around it. If you spend too long in a warm room, the molecules close to your skin surface vibrate faster because they've got more energy. If those vibrations were passed deeper into your body, they could start to disrupt the workings of your cells. So as you're sitting in the warm room, you need to give away energy. That sounds easy; water molecules evaporate easily in the warm, taking energy with them. You've got lots of water in you that could evaporate. But the water is stuck inside because you're waterproof. You need to sweat.

Your skin has a very thin layer of fat molecules just underneath the outermost skin cells, a barrier preventing any liquid moving between inside and outside. But while you're sitting in the warm room, your skin will open up tunnels through the barrier: your pores. Sweat seeps through the pores, penetrating the waterproof layer, and reaches the outside. Individual water molecules bump into each other and the warm skin surface until the most energetic are travelling fast enough to escape. One by one, they float away, leaving your skin cooler. When you are cool enough, the pores close, and you are waterproof again. Your skin isn't waterproof just to keep water out. It's also waterproof to keep water in, because the internal supply of water is limited. Water is transported round your body in your blood, the internal supply system that lets your body share out its resources. This supply system must run continuously to keep your cells alive. And we can check that it's working: we all have a pulse.

Our pulse is a three-dimensional disturbance, a travelling pressure wave that provides clues about blood flow. Our hearts are constantly squeezing the blood in their chambers, raising the fluid pressure and so forcing the blood out into our arteries. It's a powerful push, and as it comes to an end, the fluid pressure in the heart chambers drops. The forces on the blood have now been reversed, and the recently expelled blood would rush back in if it weren't for the one-way valves guarding the exit. The sudden rush of fluid backwards closes the valves, and the liquid thumps against the valve tissues as it's stopped. This thump is so strong that it pushes outwards on the tissues around it, and they push on the tissues beyond, and a pressure wave travels through the body, slightly compressing the muscle and bone in its path as it travels. It takes this pressure wave about six milliseconds to reach the outside of the body, and if you put a stethoscope or your ear up against someone's body, you can hear it. This is your heartbeat. If waves didn't travel through our tissues, we wouldn't ever be able to hear our hearts. In fact it's a double beat, with two pulses: 'lub-dub', because the heart has four valves and they close in pairs, one pair just after the other. This accidental combination of physics and physiology broadcasts the most significant sign of life throughout our bodies.

After sweating, your blood carries fewer water molecules

than before. Now, your body needs to replenish itself from outside. In order for you to take a simple drink of water, your cells need to coordinate their activities. Decisions, and the actions required to coordinate the necessary body parts to carry them out, are made first subconsciously and then consciously in the brain.

One brain cell is no use by itself. It only works because it's connected to others, and the network of connections seems to be as important as the brain cells themselves. As a decision about finding a drink emerges from the connections, the brain cells need to connect with other cells further afield. The vehicle for this internal communication is a nerve fibre, a thin strand of cell that is the body's equivalent of an electrical wire. By shunting electrically charged particles across a membrane at one end of the nerve fibre, the brain cell starts an electrical signal which ripples along the nerve fibre by playing electrical dominoes. At the end of the first nerve fibre, there's another one. The dance of the electrically charged particles sends the message across the gap, and then more electrical dominoes carry the message onwards. The message is relayed from cell to cell for the fraction of a second that it takes to reach one of the muscles in your leg. At around the same time, the messages from other nerve fibres carrying coordinated signals to other leg muscles also arrive, and your leg muscles contract to lift you off the sofa. The feel of the floor beneath your feet and the temperature change on your skin as the movement generates a slight breeze are conveyed back to your brain via yet more electrical signals.

There is a phenomenal amount of information being shunted about inside us, carried either by electrical nerve signals or by chemical messengers such as hormones. All the disparate organs and structures of a human constitute a single organism because we are connected not just by resources, but by information, vast, coordinated, overlapping streams of it. Long before the 'information age', we ourselves were information machines.

259

That information falls into two categories. The first is the travelling information: nerve signals and chemical signals that are moving right now, seeping and flashing and flowing through us. But we also carry vast quantities of stored information, the molecular library that is filed away in our DNA. In the world around us, millions of similar atoms clump together to form large agglomerations of glass or sugar or water. But in the giant molecule that is a DNA strand, each minute atom sits in its pre-scribed place, and the precise placing of individual atoms of different types gives our bodies an alphabet. A piece of the cell's molecular machinery can walk along the strand, reading the genetic alphabet of A, T, C and G, and use that information to build proteins or regulate the activity of the cell. We have to be gigantic compared with atoms because each factory-like cell has to contain so much.

Our bodies are immense machines; even a single cell might contain a billion molecules, and there are around 10 million million (10^{13}) cells in our body. We need impressive signalling and transport systems to coordinate all these constituent parts, and that coordination takes time. No human has 'lightning reac-tions', because the cost of our wonderful complexity is the huge amount of time it takes us to get anything done. The shortest time that we can appreciate is approximately the blink of an eye (about a third of a second), but in that time millions of proteins have been built inside us and billions of ions have diffused across our nerve synapses, while the simpler world outside our bodies has just been getting on with things.

Our internal information engine carries on churning as we walk from one room to another. But this gigantic system needs information about what's around it. Just at this moment, we need to find water. We have sensors built in, body parts that change in response to the environment and share that information with our brain. The sense that we're most aware of is probably our sight.

We live submerged in light, but our body keeps most of it out. That sea of light carries information about the world, because light's nature offers clues to its origins, but most of this information goes straight past us. A tiny fraction of this luminous cornucopia falls on to the pupils of our eyes, two circles at most a few millimetres in diameter. A small subset of what arrives at the pupils, the visible light rays, is allowed in. From this tiny sample comes all of the visual richness that we take for granted. As they cross the boundary, these light waves must be marshalled so that the information can be harvested. Our windows on the world are guarded by soft, transparent lenses which slow the light to 60 per cent of its speed in air. As these light rays are slowed down, they swerve, and the lens shape is tweaked by tiny muscles to make sure that all the rays from a single object out there, outside the body, will meet again at the back of the eye. This selection process is astonishing. We assume we see all there is, but really we sample only the tiniest fraction of what is out there to build our picture.

The light rays hitting our retina may have travelled from the Moon or from our fingers, but they have the same effect. A single photon is absorbed by a single opsin molecule, twisting the molecule around to start a chain of dominoes that sends an electronic signal into our control systems. As our thirsty body walks into the kitchen, photons that have bounced off a sink, a tap and a kettle stream into our eyes, and our brain processes that information in the blink of an eye to tell us what to pick up first. If it's slightly dark in the kitchen, we turn on a lightbulb, releasing a fountain of light waves. They radiate outwards, and as soon as their journey starts they're being modified by the world, reflected, refracted and absorbed until perhaps our eyes pick up what's left. And it isn't just light that's flowing around us.

Humans are sociable animals. We hold our social networks together via communication, sending and receiving signals from others. Our voice is one of our most special features, a flexible musical instrument capable of producing and shaping sound waves which are then broadcast through our surroundings. No Brit could imagine making a hot drink without asking others in the room whether they wanted to share the experience, and we ask with sound. The others in the room pick up the signal with their ears, and hearing the question will trigger a new flow of information inside their bodies, splitting and rejoining and gathering meaning until their nerve fibres instruct their vocal muscles to provide the appropriate response. Once the message has come back to us, we alter the world appropriately, rearranging the ceramics and metals in front of us.

Many different atoms make up our bodies, and wonderful though the variety is, there are limits to what we can do directly because of the way those atoms are arranged. But humans are experts at manipulating the world to produce tools that can do what we can't. We can't hold water in our hands as it boils, but a steel kettle can. We can't turn a part of ourselves into an air-proof container for dried leaves, but a glass jar will do that job. We don't have claws or a shell or tusks, but we can make knives and clothes and can-openers. A ceramic container can hold a hot drink for us without transmitting the heat energy to our vulnerable and sensitive fingers. Metals, plastics, glass and ceramics are all our proxies in the world, helped by the materials with biological origins: wood, paper and leather.

The kettle has held the water molecules while giving them energy in the form of vibrations at the tiniest level. Now they're jostling far more quickly than before, and we transfer them to their new ceramic home. Frustratingly, we can only see a hint of the splash of milk that bounces back up just after we add it to the drink. It's right there, right in front of you in plain sight, but it's virtually invisible because we just can't process the signals fast enough. You can no longer see the bottom of the cup: the liquid that had been partly transparent is now opaque because light is bouncing off millions of tiny fat droplets.

As we're manipulating the world around us, we take for granted that we are stuck to the floor by a force that's manageable because our bodies have evolved to cope with it. If the Earth's gravity were stronger, we'd probably need sturdier legs and we might find bipedal life difficult. If gravity were weaker, we might have evolved to be taller, but life would be slower because everything would take longer to fall. As we lift one leg to step, we're relying on the pull of gravity to make us fall forwards. We pivot around our stationary foot and by the time we stop the fall with the stepping foot, our whole body has moved forwards. Walking doesn't work without gravity, and our bodies have evolved to suit the Earth's gravity. We're just the right size and shape for bipedal walking to work for us. As we pick up the drink and move to the door, we're using our own bodies as an upside-down pendulum, swinging each leg forwards as we pivot about our other foot and hip. The rhythm in our walking caused by that regular swinging affects the liquid in the cup, forcing it to slosh with the same rhythm.

As we walk, we use fluid tucked inside our skulls to help with our balance. The fluid sloshing around deep inside the tiny cavity of our inner ear keeps going when we stop and lags behind when we start. Sensors on the walls of that cavity feed that information into the giant connected network of our brain, helping with the decisions about which muscle to move next.

On this occasion, we reach a door, push it open with our free hand, and step outside.

Earth

Outside, we can look sideways through the invisible atmosphere, at the rest of the world. Our planet is a system made of five interacting components: rocks, atmosphere, oceans, ice and life. Each one has its own rhythm and dynamics, but the sumptuous variety that we see on Earth is the result of the timeless dance that connects them. The same forces drive them all, and there are similarities in surprising places. As we look out through the invisible molecules that fill the sky, packets of air are shifting according to their buoyancy. Air that has been warmed by the building we've just stepped out of is rising because it's less dense than the air around it. Columns of rising air from warm ground could be a few kilometres tall, taking perhaps five minutes or so to rise each kilometre. Cooler, denser air is flowing in beneath to take its place, hauled downwards by Earth's gravity. These patterns of convection stretch out across the landscape we're looking at. Air is never completely still.

If we were looking sideways across the surface of the deep ocean, our view might encompass similar buoyancy flows, also invisible. Cold, salty water in the North Atlantic sinks downwards towards the centre of the Earth, just like the cooler, denser air. Once it reaches the ocean floor, it flows sideways across the seabed until it warms or mixes with less salty water and floats back to the surface. In the sky, one cycle of floating up and sinking down might take a few hours. In the ocean, one cycle might take four thousand years, and the water will be carried around half the globe as it happens.

And then, down below our feet right now, the rocks themselves are also moving. The Earth's mantle makes up most of the planet, a thick layer between the outer core and the thin crust floating on top. It's liquid but viscous, sluggish and slow. This melt is being heated, both by the Earth's hot core and by the slow decay of radioactive elements buried deep inside it. This shunting of energy around the deep rocks is happening now, beneath all of us. As the hot mantle rock becomes buoyant, it floats upwards and cooler rock sinks down to take its place. But molten rock at these temperatures and pressures takes time to move. Deep below us, a mantle plume might take a year to float upwards by 2 cm. A full cycle from the bottom to the top and back again might take 50 million years. But the centre of the Earth bows to the same physics as the atmosphere and the ocean, continuously shifting heat from within to without.

A vast amount of heat energy is continuously moving outwards from the centre of the Earth, but it's utterly insignificant compared to the amount of light energy from the Sun striking our planet. And in almost every environment on Earth, tucked away in corners or dominating the landscape, there is green. It might be a furtive veneer of moss on a brick wall or the luxuriant biological architecture of a rainforest, but plants are everywhere. Each leaf is the support structure for layers of chlorophyll-stuffed cells, each one a tiny molecular factory turning sunlight and carbon dioxide into sugar and oxygen. A fraction of the energy in the flood of light washing over each leaf is captured and stashed away as sugar: fuel for the future. Even on the calmest of sunny days, in a field where everything looks still and unchanging, the plants are busy. One molecule at a time, they are producing the oxygen that we breathe, enough to keep all the other living things on Earth alive, enough to maintain an atmosphere that is 21 per cent oxygen. These tiny molecular machines are continuously remaking a fifth of our entire planet's atmosphere. As we look sideways through the air, we're looking through the jostling molecular output of millions of ferns, trees, algae, grasses and more, produced over thousands of years: the bounty of a green army.

From our position on the ground, just outside our home, we can see only a small fraction of our planet. Suppose we can levitate, and we'll see far more. As we rise through the atmosphere, the air molecules spread out. Gravity is pulling them downwards, and it can only hold a very thin layer to the Earth's surface. As we rise past the top of the largest thunderstorm, approximately 20 km up, 90 per cent of the molecules in the atmosphere are beneath us. The deepest point in the ocean is 11 km below sea level, and below that, there is dense rock for about 6,360 km before you get to the centre. Without a rocket, we humans are confined to a vertical range of a measly 30 km, playing on the edge of the giant planet we call home. A layer of paint coating a ping-pong ball has the same thickness relative to the sphere it covers.

At a height of 100 km, we are officially at the boundary between Earth and outer space, and we can see the globe rolling past beneath us – green, brown, white and blue, spinning in the blackness of space. From up here, the scale of the ocean is shocking: a planet-sized shell made of one simple molecule repeated over and over again. Water is the canvas for life, but only in the Goldilocks zone,[1] the energy range within which the molecules move about as a liquid. Give those molecules extra energy, and their vibrations will shake apart any complex molecules that they house. More energy still, and they will float away as a gas, useless for protecting fragile life. At the lower end of the Goldilocks range, as you reduce the energy, the vibrations slow until the molecules must slot themselves into an ice lattice. Immobility like that is the enemy of life. Even the process of building these inflexible ice crystals can burst any living cell that contains them. Our planet is special not just because it has water, but because that water is mostly liquid. From our vantage point here on the edge of space, Earth's most precious asset dominates the view.

Perhaps, down there, as the Pacific Ocean glides past, a blue whale is making sound waves, calling into the gloom. If we could watch that sound travelling beneath the ocean surface, we would see it travelling outwards like ripples on a pond, taking an hour to reach California from Hawaii. But the sound is hidden in the water, and no evidence of it is visible from up here. The oceans are filled with sound, overlapping pressure oscillations pulsating outwards from breaking waves, ships and dolphins. The deep rumblings of Antarctic ice can travel underwater for thousands

[1] Not too hot, not too cold, but just right.

of miles. From our viewpoint on the edge of space, you would never know that any of it was there.

Everything on the planet is spinning, travelling once around the Earth's axis every day. As winds travel across the spinning surface, they tend to keep going in a straight line, although friction with the ground and confinement by the air around them constrains their path. From up here, we can see that the winds in the northern hemisphere tend to turn to the right, relative to the ground, as they carry on in spite of the spin of the Earth. So weather, especially the weather further away from the equator, spins. Hurricanes rotate, and so do the smaller storms that we can see rolling across the oceans. The eye of the storm is the hub of each wheel, and each wheel must spin because the Earth spins.

Over Antarctica, thick snow clouds are gathering. Inside each one, billions of individual water molecules exist as a gas, jiggling around with the oxygen and nitrogen. But as the cloud cools, they are giving up their energy and slowing down. When the most sluggish molecules bump into a nascent ice crystal, they lock on, each in a fixed place in the ice lattice. As the snowflake is buffeted up and down inside the cloud, the molecules on all six sides of the original crystal find themselves in the same conditions, and stick in the same way. Molecule by molecule, a symmetrical snow crystal is built. After hours of slow growth, the crystal is large enough for gravity to win the battle and it tumbles from the bottom of the cloud. Below is the Antarctic ice sheet, the largest agglomeration of ice on Earth, stretching sideways for thousands of kilometres and down for thicknesses of up to 4.8 km. The accumulation of ice is so heavy that the continent itself has been pressed downwards under the additional weight. But every molecule of that white expanse fell in a snowflake, and the pile of snowflakes has been growing for a long time. Some of the water here has been frozen for a million years. In that time, the molecules have vibrated about their fixed crystal lattice location continuously, but never fast enough to become a liquid again. In contrast, the molecules being pushed out of Hawaii's volcanoes as lava are only just dropping below 600°C for the first time since the Earth was formed, 4.5 billion years ago.

At the heart of Earth's outer engine is the energy supply from the Sun. As it heats up the rocks, ocean or atmosphere, or as it fuels sugar production in plants, it's pushing the engine away from equilibrium. As long as there's an imbalance in the distribution of energy, there is always the potential for things to change. The movement energy of falling rain can erode mountains as it splashes down on bare rock. The vast excess of heat energy at the equator drives tropical storms, battering palm trees, redistributing water from sea level to high mountains, and sending waves crashing on to beaches. The energy stored in a plant will be used to build branches, leaves, fruit and seeds, eventually running out of usefulness as low-level heat. Only the seed will be left, a package of genetic information destined to restart the cycle with new energy from the Sun's fountain of light. Our planet lives because of the constant injection of energy from above, feeding the engine and preventing Earth winding down into stable, unchanging equilibrium. From up here on the edge of space, we can't see the tiny details, but we can see the big picture: energy flows on to Earth from the Sun, trickles down through the ocean, the atmosphere and life, and eventually carries on out into space as the planet radiates heat away. The same amount of energy goes in and comes out. But the Earth is a gigantic dam in the energy flow, storing and using this precious resource in myriad ways before it's released to the universe.

As we drift back down to ground level, a beach now looks like a process instead of a place, a patchwork of timescales and size scales. The ocean waves are carrying energy from storms far out at sea. As they break on the beach, they rattle the sand and rocks, grinding them together. One speck at a time, the stone is chipped away, each pebble sculpted by millions of random collisions. It takes a millisecond to remove one minuscule chip, but years of slow attrition to make the pebble smooth. In geological time, a beach is temporary. It will last only if the supply of new pebbles and sand is greater than the loss as they wash out to sea. Over months and years, sand will shift into the sea and back out again in response to the ocean. We love our tidal beaches precisely because we can see the ebb and flow reshaping the sand twice a day; it's as though the slate is wiped clean, and we find the simplicity of the newly smoothed sand satisfying. But this daily remodelling hides the decadal shifts as our coastline grows and shrinks in front of us. The life in the rock pools thrives on change, adapted to periods of being high and dry alternating with spells of complete submersion. Though a casual glance at a rock pool can give the impression of a museum exhibit behind glass, in every pool there is a fierce battle for resources going on. The resources on offer are all ultimately very simple: access to the drips of energy oozing through the Earth's system or the chance to gather the molecular building blocks needed to construct a life. More than anywhere else, a beach exemplifies the transience of life. When the energy and nutrients are available to support life, rock pools flourish. During the barren periods, life will be found elsewhere. Species evolve by altering their use of the physical toolbox available to them one genetic mutation at a time. Whether they're harvesting energy, moving around, communicating or reproducing, they are all just using the same principles in different ways.

Energy passes through, but the Earth itself is constantly re-cycled. Almost all the aluminium, carbon and gold that make up our planet has been here for billions of years, shifting from one form to another. It might seem that after this long, these different substances should all be jumbled up, mixed together in a giant planetary soup. But the physical and chemical processes around us are continually sorting the pile, so that pockets of similar atoms group together. Gravity allows liquids to drain through porous solids, so that water soaks into the soil and joins huge underground aquifers while the soil stays where it is. When vast blooms of tiny calcium-based marine creatures live and then die at the ocean surface, it's gravity that coaxes them downwards so that they drift towards the ocean floor. The vast marine cemeteries that sometimes form as a result in shallow seas compress, shift, and become distinctive white limestone. Salt deposits are formed because water molecules will evaporate easily to become a gas when they are given more energy, but salts won't. The lava produced at volcanic mid-ocean ridges is far more dense than water, so it stays on the ocean floor, building new crust. And life itself is constantly plucking materials from the world around it, reshaping and reorganizing them, and then leaving the detritus to be reused when it dies.

On a dark night, looking up at the sky, we see waves that have travelled across our solar system or our galaxy or our universe to reach our eyes. For millennia, light waves were our only connection to the rest of the universe, the only reason we knew that there was anything else out there. A couple of decades ago, we started to observe the thin streams of matter that reach us: neutrinos and cosmic rays. And then gravitational waves came along, only the third way that we have of touching the rest of the universe. In February 2016, it was finally confirmed that catastrophic astronomical events like the merging of black holes also send out waves, ripples in space itself. Gravitational waves have been passing through all of us our entire lives, and we're finally about to discover what we've been missing out on. The light and gravitational waves whooshing past our planet weave a rich tapestry that lets us map out our universe and then add an arrow labelled: 'We are here.'

But on an average day on Earth, there are more immediate considerations. Standing outside our home and watching the world go by is a reminder of the gigantic system that we're part of. We are a small sliver of the life that keeps the system running in its current configuration. When *Homo sapiens* first emerged, each human only had two life-support systems: a body and a planet. But now there is a third.

This planet has been altered by many species, but only in the past few thousand years has a single species knowingly rebuilt its environment to suit itself. It is almost a single organism now, a sprawling planet-sized web of interconnections between individual consciousnesses. Each individual is almost entirely dependent on others in the system for survival, but still has its own contribution to make. An understanding of the laws of physics is one of the pillars holding up our society, and we could not manage our transport, resource management, communication or decision-making without it. Science and technology make possible the greatest ever collective human achievement: our civilization.

Civilization

A candle and a book. Portable energy and portable information, available on demand but with the potential to last for centuries. These are the threads that stitch individual human lives together to build something much bigger: a cooperative society that is always building on the work of the previous generation. Energy must keep flowing through our civilization, so the candle can be stored almost indefinitely but can only be used once. Knowledge accumulates, so one book may stimulate many minds. There were candles and books two thousand years ago and there are still candles and books now. They are simple technologies, but they work. We have built the modern world by storing energy and sharing information about what to do with it.

We associate civilizations with cities, but they are always founded in the fields. It takes energy to build, to explore, to try and fail and try again, and so humans had to marshal the plants to harvest solar energy in order to fuel their efforts. Humans can move soil and water and seeds, but we need plants to convert light waves into sugar. We learned how to put our own green dam in place to divert a tiny part of the torrent of solar energy, and we reaped the rewards. As it trickled through Earth's system, that temporarily diverted energy fed us, fed our animals, and gave us the capacity to alter our world.

We think of ourselves as living in a modern society, but that's only partly true. We rely on infrastructure built by previous generations, sometimes decades ago, sometimes centuries ago and sometimes millennia ago. Those roads and buildings and canals are still useful because they are the conduits that connect the distant and disparate parts of our society. Cooperation and trade bring enormous benefits, and these networks give each individual access to far more than their own solitary strength and intelligence could bring them.

A city is a forest of buildings, each with a different function and a different design. But underneath them all runs a huge web of thick copper cables. The copper tendrils branch as they run into individual buildings and then branch again and again, hidden in the walls and floors until the tip of one offshoot finally becomes visible at each power socket. As soon as something is plugged in, a loop is completed and electrons are free to shuffle around it, linking the outwards branching structure with the merging return structure. If you could see only the cables and not the city, you'd see the arteries of modern life, feeding us with energy from the massive power generation facilities elsewhere. The network extends across each country, a metal network of linked routes, connecting up the huge range of energy sources capable of collectively feeding the monster. We are surrounded by drifting electrons doing our bidding.

Overlaid on top of the power network there are other networks, also reaching up into the buildings and into our lives. Earth has its own planet-sized water cycle, linking oceans with rain and rivers and aquifers. Energy from the Sun provides the energy to evaporate water, to move it through our atmosphere, and to deposit it somewhere else. We humans build local diversions, funnelling water out of the natural cycle and pumping it through our civilization before releasing it back to the world. Rain that has collected in a reservoir is held back, prevented from following gravity's call directly into rivers and then the ocean. Shuffling electrons provide power to pumps that send water through pipes nearly a metre in diameter, out from the reservoir, branching and branching as they travel out along our roads, into our buildings and finally up to our taps. When we have used it, it travels back through drains and sewers, through pipes gradually increasing in size as they join together on their way to a water treatment plant or a river. When we turn on a tap, we see the tip of the network, a small link in a giant loop. Then the water runs away, out of our sight, back into the concealed tunnels. Gravity keeps it in check; as long as we do the initial work to lift it up, putting in the energy to push the water away from equilibrium, gravity will always take over to guide the onward flow down again. The drain is just the place where the resistance against gravity temporarily disappears.

A city is the place where these networks and others are all compressed together because in these places the humans are compressed together, relying on the networks to live. There are other networks overlaid on the familiar city scene: systems of food distribution, human transport and trade to share resources. And these are just the ones that are visible if you know where to look.

Fire was the start of the human adventure with artificial light. Instead of relying on light waves from the Sun, we learned to make our own. Candles meant that we could see even when our side of the Earth had rotated around to face away from the Sun. A hundred and fifty years ago, a city at night was lit up by the light waves given out by burning candles, wood, coal and oil.

Today, the sky is filled by light that we can't see, shining all day and all night. If we could see radio waves, we'd see that our planet hasn't been dark at those wavelengths for a century. But these new waves are more than illumination. Radio waves, television broadcasts, wifi and phone signals form a tightly coordinated network of information, constantly rippling through our surroundings and ourselves. Anyone standing in the midst of our civilization with an electronic device that can listen in to precisely the right type of wave immediately has access to visual news broadcasts, shipping forecasts, reality TV shows, air traffic control, ham radio and the voices of friends and family. The waves are streaming around us all the time, and the wonder of the modern world is that it's so easy to listen in and to contribute. The flow of information ties our world together. Farmers can plan harvests based on what the supermarkets want this week. News of natural disasters touches the rest of the planet in real time. Planes can reroute to avoid bad weather up ahead. A trip to the shops can be postponed because the rainclouds will arrive overhead in ten minutes' time. The system works because the waves are coordinated by humans cooperating with each other, because our species agreed on global rules for some waves and national rules for others. For most of human history, there were waves but no network. In only the past five generations, humans have constructed the wave-based information network that we now all find indispensable.

In the past, humans have been limited geographically by heat, cold or the lack of resources. If the molecules around us have too little heat energy or too much, the molecules that make up our own bodies follow suit. If the careful balance between molecular activity and stasis in our bodies is lost, we start to suffer. But these geographical limits have been almost entirely lifted. We construct buildings, walkways, vehicles and barriers and alter the inside of each structure so that it has the right energy level for us to be comfortable. Air-conditioning in Dubai and central heating in Alaska give us habitable bubbles where none existed before. We forget about the inconvenience of the real world, and take our protective bubbles for granted. Human habitations on other planets are still a long way off, but humans have developed some of the technologies required to make more of our own planet habitable. The principle is the same: manipulating an environment until it fits our strict conditions for survival. The supply of water, molecular building blocks and energy must all be just right. When we have built one bubble, we build another, creeping across our planet and extending our survival networks wherever we go.

As our civilization grows, we face challenges. The larger the human population is, the more resources and space are required to support us. We have discovered that our use of the fuels that powered the Industrial Revolution and the dramatic growth of the developed world comes at a cost. At the same time as humans were growing plants to harvest the Sun's energy, building a green energy reservoir that could be manipulated as required, most of our energy came from another source. Earth already had an energy reservoir formed from the torrent of solar energy, one that had collected over hundreds of millions of years, and we supped from that. Over eons, a small fraction of the plants that have trapped the Sun's energy were themselves trapped, buried and compressed deep underground. The slow accumulation of captured solar energy built up a huge subterranean store, stashed away safely as the flow of solar energy to and from the planet continued at the surface. We call these ancient energy stores fossil fuels, and the energy is easily released and put to work. Using the energy itself isn't a problem; it's just stored solar energy which is finally being released back out into the universe. Knowing what to do with the packaging is a nightmare. Plants take in carbon dioxide to grow, and as the energy from their fuel is released, carbon dioxide is also re-formed and given back to the atmosphere. These individual gas molecules drift out into the air, changing how waves pass through the atmosphere. The consequence is that the planet overall becomes a slightly bigger reservoir for the Sun's energy. After burning through the energy stores of millions of years, humans have heated the planet up slightly. Learning to deal with the new equilibrium state of our planet will take considerable ingenuity.

But humans are ingenious. Our understanding of science, medicine, engineering and our own culture is now available to be plucked from the network of invisible waves around us. Every time we use anything from that information network, we are benefiting from the effort put in by generations of other humans.

One of the greatest advances has come from discovering how much space there is to work with if you play on size scales other than our own. The human body and the structures that fit it aren't changing size – we consist of a hugely complex system and we need this much space to contain it. The size of beds, tables, chairs and food won't change, because we each live in *this* body. But as we learn to manipulate the world of the small, and shrink our view accordingly, humans are also learning to construct enormous factories that are too small for our bodies to see. The time taken to get things done shrinks as the size shrinks, so billions of processes can be carried out in each second. Electricity flows easily on those tiny scales. A computer is just an electronic adding machine made out of nanoscale components. Computers seem small to us, but compared with the atoms they're constructed from, they're vast architectural marvels, with function built into their form. Astonishment at the seeming magic of a computer is really the shock of accepting that things can happen on other time and size scales. Even now, these tiny giant adding-up factories are becoming essential tools for controlling our world, and they will become more deeply integrated into our civilization as time goes on. A more crowded civilization requires more efficient decisions, faster decision-making, and a faster flow of information to coordinate the delicate cogs of the system. Using size scales other than our own makes this possible.

Our species is currently confined to this planet and its near environs, but we have looked outward at the stars for generations. Now, for the first time in the history of human civilization, we are also looking back at ourselves. Earth observation satellites and communication satellites swarm around our planet, connecting us to each other and allowing us to watch the globe roll by beneath them. From up there, the imprint of our civilization is visible: bright city lights at night, warm air around cities in cold places, the changed colour of the land from agriculture. Just one of those orbiting objects is a bubble suitable for humans: the International Space Station. Our civilization does extend into space – just. A maximum of ten people at a time can represent the rest of humanity up there, orbiting the Earth once every ninety-two minutes. The men and women who have seen their planet from orbit understand that they share a perspective on our civilization that they will never completely convey to the rest of humanity. But to their enormous credit, they try.

Above the satellites, and well outside the magnetic shield that protects our planet from cosmic rays, the signs of our civilization dwindle. Out here in space, there is no up and down. A pendulum clock won't tick because there is no gravitational pull on the pendulum. The simplicity of things here means that everything happens either exceptionally fast by human standards, or exceptionally slowly. Rapid nuclear reactions power the Sun, but the Sun changes only slowly over billions of years. Tiny atoms interact, and the outcomes are the size of a planet or a moon or a solar system. Our messy, complex civilization on our messy, complex world sits in the middle of the size and time scales.

We are an exception in the known universe.

Humans look out at space, and maybe something out there in space is looking back. Light is still our main connection to everything that isn't our planet, and the molecular shifts caused when

277

starlight hits our retina link us to the rest of the universe. Here we are, a beautiful, complicated, sentient layer, a thin coating on a small rocky planet, living on the boundary between the cosmos and the Earth. Here we are, a product of our three interlinked life-support systems, shaped by the physics of the universe.

Here I am, standing outside my home, looking out at the sky as the clouds gather and hide the rest of the universe from my gaze. Here I am, a modern human with a mug made from the Earth, thinking about the complexities of the universe because I can. The patterns are all around me, and I can touch them for myself. I look into my teacup and see the swirling liquid. And then I look again and see something different. Reflected from the liquid surface is a similarly bright, beautiful and fascinating pattern, an image of the sky above my head. Right there, in my teacup, I can see the storm.

References

Chapter 1: Popcorn and rockets

Ian Inkster, *History of Technology*, vol. 25 (London, Bloomsbury, 2010), p. 143

'Elephant anatomy: respiratory system', Elephants Forever, http://www.elephantsforever.co.za/elephants-respiratory-system.html#.VrSVgfHdhO8

'Elephant anatomy', Animal Corner, https://animalcorner.co.uk/elephant-anatomy/#trunks

'The trunk', Elephant Information Repository, http://elephant.elehost.com/About_Elephants/Anatomy/The_Trunk/the_trunk.html

John H. Lienhard, *How Invention Begins: Echoes of Old Voices in the Rise of New Machines* (New York, Oxford University Press, 2006)

'Magdeburger Halbkugeln mit Luftpumpe von Otto von Guericke', Deutsches Museum, http://www.deutsches-museum.de/sammlungen/meisterwerke/meisterwerke-i/halbkugel/?sword_list[]=magdeburg&no_cache=1

'Bluebell Railway: preserved steam trains running through the heart of Sussex', http://www.bluebell-railway.co.uk/

'Rocket post: that's one small step for mail . . .', *Post&Parcel*, http://postandparcel.info/33442/in-depth/rocket-post-that%E2%80%99s-one-small-step-for-mail%E2%80%A6/

'Rocket post reality', Isle of Harris website, http://www.isleofharris.com/discover-harris/past-and-present/rocket-post-reality

Christopher Turner, 'Letter bombs', *Cabinet Magazine*, no. 23, 2006

References

'A sketch diagram of Zucker's rocket as used on Scarp, July 1934 (POST 33/5130)', Bristol Postal Museum and Archive

Chapter 2: What goes up must come down

D. Driss-Ecole, A. Lefranc and G. Perbal, 'A polarized cell: the root-statocyte', *Physiologia Plantarum*, 118 (3), July 2003, pp. 305–12

George Smith, 'Newton's *Philosophiae Naturalis Principia Mathematica*', in Edward N. Zalta, ed., *Stanford Encyclopedia of Philosophy*, Winter 2008 edn, http://plato.stanford.edu/archives/win2008/entries/newton-principia/

Celia K. Churchill, Diarmaid Ó Foighil, Ellen E. Strong and Adriaan Gittenberger, 'Females floated first in bubble-rafting snails', *Current Biology*, 21 (19), Oct. 2011, pp. R802–R803, http://dx.doi.org/10.1016/j.cub.2011.08.011

Zixue Su, Wuzong Zhou and Yang Zhang, 'New insight into the soot nanoparticles in a candle flame', *Chemical Communications*, 47 (16), March 2011, pp. 4700–2, http://dx.doi.org/10.1039/C0CC05785A

Chapter 3: Small is beautiful

Peter J. Yunker, Tim Still, Matthew A. Lohr and A. G. Yodh, 'Suppression of the coffee-ring effect by shape-dependent capillary interactions', *Nature*, 476, 18 Aug. 2011, pp. 308–11, http://dx.doi.org/10.1038/nature10344

Robert D. Deegan, Olgica Bakajin, Todd F. Dupont, Greb Huber, Sidney R. Nagel and Thomas A. Witten, 'Capillary flow as the cause of ring stains from dried liquid drops', *Nature*, 389, 23 Oct. 1997, pp. 827–9, http://dx.doi.org/10.1038/39827

The whole of the *Micrographia* is online here: https://ebooks.adelaide.edu.au/h/hooke/robert/micrographia/contents.html

'Homogenization of milk and milk products', University of Guelph Food Academy, https://www.uoguelph.ca/foodscience/book-page/homogenization-milk-and-milk-products

'Blue tits and milk bottle tops', *British Bird Lovers*, http://www.britishbirdlovers.co.uk/articles/blue-tits-and-milk-bottle-tops

References

Rolf Jost, 'Milk and dairy products', in *Ullman's Encyclopedia of Industrial Chemistry* (New York and Chichester, Wiley, 2007), http://dx.doi.org/10.1002/14356007.a16_589.pub3

Aaron Fernstrom and Michael Goldblatt, 'Aerobiology and its role in the transmission of infectious diseases', *Journal of Pathogens*, 2013, article ID 493960, http://dx.doi.org/10.1155/2013/493960

'Ebola in the air: what science says about how the virus spreads', *npr*, http://www.npr.org/sections/goatsandsoda/2014/12/01/364749313/ebola-in-the-air-what-science-says-about-how-the-virus-spreads

Kevin Loria, 'Why Ebola probably won't go airborne', *Business Insider*, 6 Oct. 2014, http://www.businessinsider.com/will-ebola-go-airborne-2014-10?IR=T

N. I. Stilianakis and Y. Drossinos, 'Dynamics of infectious disease transmission by inhalable respiratory droplets', *Journal of the Royal Society Interface*, 7 (50), 2010, pp. 1355–66, http://dx.doi.org/10.1098/rsif.2010.0026

I. Eames, J. W. Tang, Y. Li and P. Wilson, 'Airborne transmission of disease in hospitals', *Journal of the Royal Society Interface*, 6, Oct. 2009, pp. S697–S702, http://dx.doi.org/10.1098/rsif.2009.0407.focus

'TB rises in UK and London', *NHS Choices*, http://www.nhs.uk/news/2010/12December/Pages/tb-tuberculosis-cases-rise-london-uk.aspx

World Health Organization, Tuberculosis factsheet 104, 2016, http://www.who.int/mediacentre/factsheets/fs104/en/

A. Sakula, 'Robert Koch: centenary of the discovery of the tubercle bacillus, 1882', *Thorax*, 37 (4), 1982, pp. 246–51, http://dx.doi.org/10.1136/thx.37.4.246

Nobel Prize website about Robert Koch, http://www.nobelprize.org/educational/medicine/tuberculosis/readmore.html

Lydia Bourouiba, Eline Dehandschoewercker and John W. M. Bush, 'Violent expiratory events: on coughing and sneezing', *Journal of Fluid Mechanics*, 745, 2014, pp. 537–63

'Improved data reveals higher global burden of tuberculosis', World Health Organization, 22 Oct. 2014, http://www.who.int/mediacentre/news/notes/2014/global-tuberculosis-report/en/

References

Stephen McCarthy, 'Agnes Pockels', *175 faces of chemistry*, Nov. 2014, http://www.rsc.org/diversity/175-faces/all-faces/agnes-pockels

'Agnes Pockels', http://cwp.library.ucla.edu/Phase2/ Pockels,_Agnes@ 871234567.html

Agnes Pockels, 'Surface tension', *Nature*, 43, 12 March 1891, pp. 437–9

Simon Schaffer, 'A science whose business is bursting: soap bubbles as commodities in classical physics', in Lorraine Daston, ed., *Things that Talk: Object Lessons from Art and Science* (Cambridge, Mass., MIT Press, 2004)

Adam Gabbatt, 'Dripless teapots', *Guardian*, Food and drink news blog, 29 Oct. 2009, http://www.theguardian.com/lifeandstyle/ blog/2009/oct/29/teapot-drips-solution

Martin Chaplin, 'Cellulose', http://www1.lsbu.ac.uk/water/ cellulose.html

D. Klemm, B. Heublein, H-P. Fink and A. Bohn, 'Cellulose: fascinating biopolymer and sustainable raw material', *Angewandte Chemie*, international edn, 44, 2005, pp. 3358–93, http://dx.doi.org/10.1002/ anie.200460587

Alexander A. Myburg, Simcha Lev-Yadun and Ronald R. Sederoff, 'Xylem structure and function', *eLS*, Oct. 2013, http://dx.doi.org/ 10.1002/9780470015902.a0001302.pub2

Michael Tennesen, 'Clearing and present danger? Fog that nourishes California redwoods is declining', *Scientific American,* 9 Dec. 2010

James A. Johnstone and Todd E. Dawson, 'Climatic context and ecological implications of summer fog decline in the coast redwood region', *Proceedings of the National Academy of Sciences,* 107 (10), 2010, pp. 4533–8

Holly A. Ewing et al., 'Fog water and ecosystem function: heterogeneity in a California redwood forest', *Ecosystems*, 12 (3), April 2009, pp. 417–33

S. S. O. Burgess, J. Pittermann and T. E. Dawson, 'Hydraulic efficiency and safety of branch xylem increases with height in *Sequoia sempervirens* (D. Don) crowns', *Plant, Cell and Environment*, 29, 2006, pp. 229–39, http://dx.doi.org/10.1111/ j.1365-3040.2005.01415.x

References

George W. Koch, Stephen C. Sillett, Gregory M. Jennings and Stephen D. Davis, 'The limits to tree height', *Nature*, 428, 22 April 2004, pp. 851–4, http://dx.doi.org/10.1038/nature02417

Martin Canny, 'Transporting water in plants', *American Scientist*, 86 (2), 1998, p. 152, http://dx.doi.org/10.1511/1998.2.152

John Kosowatz, 'Using microfluidics to diagnose HIV', March 2012, https://www.asme.org/engineering-topics/articles/bioengineering/using-microfluidics-to-diagnose-hiv

Phil Taylor, 'Go with the flow: lab on a chip devices', 10 Oct. 2014, http://www.pmlive.com/pharma_news/go_with_the_flow_lab-on-a-chip_devices_605227

Eric K. Sackmann, Anna L. Fulton and David J. Beebe, 'The present and future role of microfluidics in biomedical research', *Nature*, 507.7491, 2014, pp. 181–9

'Low-cost diagnostics and tools for global health', Whitesides Group Research, http://gmwgroup.harvard.edu/research/index.php?page=24

Chapter 4: A moment in time

Eric Lauga and A. E. Hosoi, 'Tuning gastropod locomotion: modeling the influence of mucus rheology on the cost of crawling', *Physics of Fluids (1994–present)*, 18 (11), 2006, 113102

Janice H. Lai et al., 'The mechanics of the adhesive locomotion of terrestrial gastropods', *Journal of Experimental Biology*, 213 (22), 2010, pp. 3920–33

Mark W. Denny, 'Mechanical properties of pedal mucus and their consequences for gastropod structure and performance', *American Zoologist*, 24 (1), 1984, pp. 23–36

Neil J. Shirtcliffe, Glen McHale and Michael I. Newton, 'Wet adhesion and adhesive locomotion of snails on anti-adhesive non-wetting surfaces', *PloS one*, 7 (5), 2012, p. e36983

H. C. Mayer and R. Krechetnikov, 'Walking with coffee: why does it spill?', *Physical Review E*, 85 (4), 2012, 046117

Marc Reisner, *Cadillac Desert: The American West and its Disappearing Water*, rev. pb edn (New York, Penguin, 1993)

B. J. Frost, 'The optokinetic basis of head-bobbing in the pigeon', *Journal of Experimental Biology*, 74, 1978, pp. 187–95

'Engineering aspects of the September 19, 1985 Mexico City earthquake', NBS Building Science series 165, May 1987, http://www.nist.gov/customcf/get_pdf.cfm?pub_id=908821

Daniel Hernandez, 'The 1985 Mexico City earthquake remembered', *Los Angeles Times*, 20 Sept. 2010, http://latimesblogs.latimes.com/laplaza/2010/09/earthquake-mexico-city-1985-memorial.html

William F. Martin, Filipa L. Sousa and Nick Lane, 'Energy at life's origin', *Science*, 344 (6188), 2014, pp. 1092–3

S. Seager, 'The future of spectroscopic life detection on exoplanets', *Proceedings of the National Academy of Sciences of the United States of America*, 111 (35), 2014, pp. 12634–40, http://dx.doi.org/10.1073/pnas.1304213111

Chapter 5: Making waves

A. A. Michelson and E. W. Morley, 'On the relative motion of the Earth and of the luminiferous ether', *Sidereal Messenger*, 6, 1887, pp. 306–10, http://adsabs.harvard.edu/full/1887SidM....6..306M

Sindya N. Bhanoo, 'Silvery fish elude predators with light-bending', *New York Times*, 22 Oct. 2012, http://www.nytimes.com/2012/10/23/science/silvery-fish-elude-predators-with-sleight-of-reflection.html?_r=0

Alexis C. Madrigal, 'You're eye-to-eye with a whale in the ocean: what does it see?', *The Atlantic*, 28 March 2013, http://www.theatlantic.com/technology/archive/2013/03/youre-eye-to-eye-with-a-whale-in-the-ocean-what-does-it-see/274448/

Leo Peichl, Günther Behrmann and Ronald H. H. Kröger, 'For whales and seals the ocean is not blue: a visual pigment loss in marine mammals', *European Journal of Neuroscience*, 13 (8), 2001, pp. 1520–8

Jeffry I. Fasick et al., 'Estimated absorbance spectra of the visual pigments of the North Atlantic right whale (Eubalaena glacialis)', *Marine Mammal Science*, 27 (4), 2011, pp. E321–E331

University of Oxford, press pack for Marconi exhibition: https://www.mhs.ox.ac.uk/marconi/presspack/

Bill Kovarik, 'Radio and the *Titanic*', Revolutions in Communication, http://www.environmentalhistory.org/revcomm/features/radio-and-the-titanic/

RMS *Titanic* radio page, http://hf.ro/

Yannick Gueguen et al., 'Yes, it turns: experimental evidence of pearl rotation during its formation', *Royal Society Open Science*, 2 (7), 2015, 150144

Chapter 6: Why don't ducks get cold feet?

'Molecular dynamics: real-life applications', http://www.scienceclarified.com/everyday/Real-Life-Physics-Vol-2/Molecular-Dynamics-Real-life-applications.html

'Einstein and Brownian motion', *American Physical Society News*, 14 (2), Feb. 2005, https://www.aps.org/publications/apsnews/200502/history.cfm

'Back to basics: the science of frying', http://www.decodingdelicious.com/the-science-of-frying/

'1000 days in the ice', *National Geographic*, 2009, http://ngm.nationalgeographic.com/2009/01/nansen/sides-text/4

Jing Zhao, Sindee L. Simon and Gregory B. McKenna, 'Using 20-million-year-old amber to test the super-Arrhenius behaviour of glass-forming systems', *Nature Communications*, 4, 2013, p. 1783

Intergovernmental Panel on Climate Change, *Climate Change 2007: Working Group I: The Physical Science Basis*, IPCC Report 2007, FAQ 5.1: 'Is sea level rising?', https://www.ipcc.ch/publications_and_data/ar4/wg1/en/faq-5-1.html

Oliver Milman, 'World's oceans warming at increasingly faster rate, new study finds', http://www.theguardian.com/environment/2016/jan/18/world-oceans-warming-faster-rate-new-study-fossil-fuels

'The coldest place in the world', *NASA Science News*, 10 Dec. 2013, http://science.nasa.gov/science-news/science-at-nasa/2013/09dec_coldspot/

'Webbed wonders: waterfowl use their feet for much more than just standing and swimming', http://www.ducks.org/conservation/waterfowl-biology/webbed-wonders/page2

'Temperature regulation and behavior', https://web.stanford.edu/
group/stanfordbirds/text/essays/Temperature_Regulation.html
Barbara Krasner-Khait, 'The impact of refrigeration', http://www.
history-magazine.com/refrig.html
Simon Jol, Alex Kassianenko, Kaz Wszol and Jan Oggel, 'Issues in
time and temperature abuse of refrigerated foods', *Food Safety
Magazine*, Dec. 2005–Jan. 2006, http://www.foodsafetymagazine.
com/magazine-archive1/december-2005january-2006/issues-in-
time-and-temperature-abuse-of-refrigerated-foods/
Alexis C. Madrigal, 'A journey into our food system's refrigerated-
warehouse archipelago', *The Atlantic*, 15 July 2003, http://
www.theatlantic.com/technology/archive/2013/07/a-journey-
into-our-food-systems-refrigerated-warehouse-
archipelago/277790/

Chapter 7: Spoons, spirals and Sputnik

Hugh Gladstone, 'Making tracks: building the Olympic velodrome',
Cycling Weekly, 21 Feb. 2011, http://www.cyclingweekly.co.uk/
news/making-tracks-building-the-olympic-velodrome-53916
Rachel Thomas, 'How the velodrome found its form', *Plus
Magazine*, 22 July 2011, https://plus.maths.org/content/
how-velodrome-found-its-form
'Determination of the hematocrit value by centrifugation',
http://www.hettweb.com/docs/application/Application_Note_
Diagnostics_Hematocrit_Determination.pdf
'Astronaut training: centrifuge', *RUS Adventures*, http://
www.rusadventures.com/tour35.shtml
'Centrifuge', Yu.A. Gagarin Research and Test Cosmonaut Training
Center, http://www.gctc.su/main.php?id=131
'High-G training', https://en.wikipedia.org/wiki/High-G_training
Lisa Zyga, 'The physics of pizza-tossing', *Phys.org*, 9 April 2009,
http://phys.org/news/2009-04-physics-pizza-tossing.html
Alison Spiegel, 'Why tossing pizza dough isn't just for show', *HuffPost
Taste*, 2 March 2015, http://www.huffingtonpost.com/2015/03/02/
toss-pizza-dough_n_6770618.html

K.-C. Liu, J. Friend and L. Yeo, 'The behavior of bouncing disks and pizza tossing', *EPL* (*Europhysics Letters*), 85 (6), 26 March 2009

'International Space Station', http://www.nasa.gov/mission_pages/station/expeditions/expedition26/iss_altitude.html

Eleanor Imster and Deborah Bird, 'This date in science: launch of Sputnik', 4 Oct. 2014, http://earthsky.org/space/this-date-in-science-launch-of-sputnik-october-4-1957

Roger D. Launius, 'Sputnik and the origins of the Space Age', http://history.nasa.gov/sputnik/sputorig.html

Paul E. Chevedden, *The Invention of the Counterweight Trebuchet: A Study in Cultural Diffusion*, Dumbarton Oaks Papers No. 54, 2000, http://www.doaks.org/resources/publications/dumbarton-oaks-papers/dop54/dp54ch4.pdf

Riccardo Borghi, 'On the tumbling toast problem', *European Journal of Physics*, 33 (5), 1 Aug. 2012

R. A. J. Matthews, 'Tumbling toast, Murphy's Law and the fundamental constants', *European Journal of Physics*, 16 (4), 1995, pp. 172–76, http://dx.doi.org/10.1088/0143-0807/16/4/005

'Dizziness and vertigo', http://balanceandmobility.com/for-patients/dizziness-and-vertigo/

Steven Novella, 'Why isn't the spinning dancer dizzy?', *Neurologica*, 30 Sept. 2013, http://theness.com/neurologicablog/index.php/why-isnt-the-spinning-dancer-dizzy/

Chapter 8: When opposites attract

'One penny coin', http://www.royalmint.com/discover/uk-coins/coin-design-and-specifications/one-penny-coin

'The chaffinch', http://www.avibirds.com/euhtml/Chaffinch.html

Dominic Clarke, Heather Whitney, Gregory Sutton and Daniel Robert, 'Detection and learning of floral electric fields by bumble-bees', *Science*, 340 (6128), 5 April 2013, pp. 66–9, http:/dx.doi.org/10.1126/science.1230883

Sarah A. Corbet, James Beament and D. Eisikowitch, 'Are electrostatic forces involved in pollen transfer?', *Plant, Cell and Environment*, 5 (2), 1982, pp. 125–9

References

Ed Yong, 'Bees can sense the electric fields of flowers',
  National Geographic 'Phenomena' blog, 21 Feb. 2013,
  http://phenomena.nationalgeographic.com/2013/02/21/
  bees-can-sense-the-electric-fields-of-flowers/

John D. Pettigrew, 'Electroreception in monotremes', *Journal of
  Experimental Biology*, 202 (10), 1999, pp. 1447–54

U. Proske, J. E. Gregory and A. Iggo, 'Sensory receptors in
  monotremes', *Philosophical Transactions of the Royal Society of
  London B: Biological Sciences*, 353 (1372), 1998, pp. 1187–98

'Cathode ray tube', University of Oxford Department of Physics,
  http://www2.physics.ox.ac.uk/accelerate/resources/
  demonstrations/ cathode-ray-tube

'Non-European compasses', Royal Museums Greenwich,
  http://www.rmg.co.uk/explore/sea-and-ships/facts/
  ships-and-seafarers/the-magnetic-compass

Wynne Parry, 'Earth's magnetic field shifts, forcing airport runway
  change', *LiveScience*, 7 Jan. 2011, http://www.livescience.com/9231-
  earths-magnetic-field-shifts-forcing-airport-runway-change.html

'Wandering of the geomagnetic poles', National Centers for
  Environmental Information, National Oceanic and Atmospheric
  Administration, http://www.ngdc.noaa.gov/geomag/
  GeomagneticPoles.shtml

'Swarm reveals Earth's changing magnetism', European Space
  Agency, 19 June 2014, http://www.esa.int/Our_Activities/
  Observing_the_Earth/Swarm/
  Swarm_reveals_Earth_s_changing_magnetism

David P. Stern, 'The Great Magnet, the Earth', 20 Nov. 2003, http://
  www-spof.gsfc.nasa.gov/earthmag/demagint.htm

'Drummond Hoyle Matthews', https://www.e-education.psu.edu/
  earth520/content/l2_p11.html

F. J. Vine and D. H. Matthews, 'Magnetic anomalies over oceanic
  ridges', *Nature*, 199, 1963, pp. 947–9

Kenneth Chang, 'How plate tectonics became accepted science',
  New York Times, 15 Jan. 2011

Acknowledgements

The wonderful thing about writing these acknowledgements is that the thanks fall into two categories that overlap considerably. There are the people who provided specific book-based assistance, and there are the people who were a part of the stories I've told, those who make my life richer by sharing the adventures and providing encouragement to find more. I'm incredibly grateful for both.

My partners in exploration were Dallas Campbell, Nicki Czerska, Irena Czerski, Lewis Dartnell, Tamsin Edwards, Campbell Storey and Inca the dog. The lovely people at the Green Britain Centre (whose wind turbine I visited) were extremely hospitable and very patient with all my questions. Dr Geoff Willmott and Professor Cath Noakes were extremely helpful on the topics of microfluidic devices and airborne diseases respectively. Helle Nicholson, Phil Hector and Phil Read were kind enough to read parts of the book and contribute valuable comments. Matt Kelly deserves huge credit for providing careful feedback on the book proposal and various chapters, and I have benefited enormously from him sharing his own experiences as a writer. Matt's friendship and unfailing support have meant an enormous amount to me throughout this project. Tom Wells encouraged me to get started, and has very patiently been both sounding board and guinea pig along the way. Jem Stansfield, Alom Shaha, Gaia Vince, Alok Jha, Adam Rutherford and the many other fantastic friends I've met through the world of science have been there with encouragement and laughter throughout.

Churchill College, Cambridge, was my intellectual home for many years, and is still a home in my heart now. Between them, Churchill College and the Cavendish Laboratory gave me a thorough grounding in physics. Particular mention should go to Dr Dave Green, my Director of Studies. I hope that this book meets his exacting standards, and in particular that the number of WORDS in it makes up for the lack of DIAGRAMS. My friends from Churchill are such an important part of my life, and it's wonderful to have such fantastic and consistent companions on the adventure of life.

I arrived almost accidentally in the world of bubble physics, when Dr Grant Deane of the Scripps Institution of Oceanography took a chance on someone he had never met, and agreed to take me on as a postdoc. Grant is both an awesome human being and a rigorous and enthusiastic academic, and I am very lucky to have spent time working with him. He showed me the best that academia has to offer, and set a fabulous example of how to work to the highest standards. I can't thank him enough for that opportunity, and for his subsequent support for whatever projects I've taken on.

University College London is my academic abode now and I feel very lucky to work there. I'm based in the Department of Mechanical Engineering, and I'm extremely grateful to our Head of Department, Professor Yiannis Ventikos, for his enthusiasm when I told him I was embarking on this project. Professor Mark Miodownik is a reliable fountain of energy and warmth and solid advice and friendship, and I am hugely in his debt for helping me find such a fabulous academic home.

My literary agent, Will Francis, encouraged me to write a book, was almost unbelievably patient until it was the right time, and has been a brilliant source of support and advice along the way. Susanna Wadeson at Transworld has been a steady hand on the tiller of the whole project, and I am extremely grateful to her for her insight and her honesty.

My family are a fabulous bunch, reliably curious about the world, forever supportive, always up for trying things out and bonkers in the best possible way. Everything I've done rests on the foundation that they provide. My sister Irena is amazing, and she and Malcolm are possibly the most hospitable and loveliest people I've ever met. Listening to my Nana, Pat Jolly, and my Aunt Kath and my Mum share the stories of early televisions and the mysteries of a line output transformer was something that I should have done properly years earlier. The most important thanks go to my parents, Jan and Susan. They showed us how to explore the world and asked just that we did our best. I love them both and I can't thank them enough.

Index

AC (alternating current), 236–7
AC/DC adaptor, 237
Adams, Douglas, 204
adiabatic heating, 23
air: atoms, 13–14; baking, 17–20; breathing, 256–7; bubbles, 76–9; buoyancy, 57–9, 62; buoyancy flows, 264; convection currents, 61, 264; dogs panting, 114; floating particles, 75–6; hot, 60–1; katabatic winds, 21, 22–3; molecules, *see* air molecules; pockets, 40–1, 85–6; pressure, 12, 23–6, 28, 40; resist-ance, 43–4, 209, 222; sucking, 26–8; swirls of warm and cold, 1–2; thunderstorm, 131–3; viscosity, 73, 74; water molecules, 161–2, 221; weather, 36–7; whales, 14–15
air conditioning, 274
air molecules: air pressure, 24–6; Antarctica, 21; atmosphere, 265; bees flying, 222; breathing, 256–7; eggs, 58; elephants drinking, 27–8; katabatic winds, 23; speed, 23; thunderstorms, 36–7; vacuum pump, 24–5
alcohol molecules, 164, 175
Amundsen, Roald, 20, 168
Antarctica: Andes, 45; ice sheet, 267; katabatic winds, 21, 22;

mid-Atlantic ridge, 250; polar explorers, 21, 22, 168; sounds of ice, 266–7; thermohaline circulation, 63–4; winter temperature, 176–7
anthocyanins, 8–9
Archimedes, 56
Arctic Ocean, 165–8, 247
Aristotle, 24
Asimov, Isaac, 157
astronauts, 192–3, 205–6
Atlantic Ocean: buoyancy flows, 264; experiments at sea, 228, 231; mid-Atlantic ridge, 250–1; thermo-haline circulation, 63–4; *Titanic* radio messages, 139–40, 150; weather, 1; width, 252
atoms, 13–14; Brownian motion, 158–9; conductivity, 181–2; crys-tals, 156–7, 169–70; DNA, 260; existence, 14, 69, 157–8, 182; glass, 171–5; ice crystals, 176–7; ice cubes, 169, 184; ions, 155n; magnetic fields, 217; nucleus, 13, 219; size, 69

baking: carrot cake, 49–50; focaccia, 17–20
balance: centre of mass, 207; equilib-rium, 105–6; in orbit, 205; terminal velocity, 44; tightrope walking, 54–5; Tower Bridge, 52–3; whale, 15

293

Index

Index

Whitesides, George, 92
wildfires, 22–3
Wilson, Robert, 145n
wind turbines, 253–4
windows, glass, 173
winds: buildings vibrating, 115; katabatic, 21–3; monsoon, 159–60; ocean currents, 63–4; spinning, 267; storms, 37, 123–4
'wise women', 8, 9

Zucker, Gerhard, 33–5

About the author

Helen Czerski is a lecturer in the Department of Mechanical Engineering at University College London. She is a physicist and studies the bubbles underneath breaking waves out in the open ocean to understand their effects on weather and climate.

Helen regularly presents BBC programmes on physics, the ocean and the atmosphere – recent series include *Colour: the spectrum of science*, *Orbit*, *Operation Iceberg*, *Supersenses* and *Dara O'Briain's Science Club* – as well as programmes on bubbles, the Sun and our weather. She is also a columnist for *Focus* magazine, was shortlisted for PPA columnist of the year in 2014, and has written numerous articles for the *Guardian*.